Living the Good Life in the City

Living the Good Life in the City

A JOURNEY TO SELF-SUFFICIENCY

Sara Ward

Pimpernel
Press ltd
www.pimpernelpress.com

CONTENTS

WELCOME TO HEN CORNER

'We're making cider,' said Andy, my husband, as we looked at the ripening apples on our newly acquired trees. Our journey into smallholding had begun with growing fruit and vegetables in our garden a few doors along the terrace from Hen Corner, but we were not there long enough to see the plum and apple trees that we'd planted start to fruit.

Then we bought the corner house, where the larger garden was already home to seven fruit trees and had a designated growing space for a vegetable plot. Two of the trees, in the centre of the lawn, may have been retained from the orchard that thrived on the site before the houses were built, which could make them around 150 years old. They are apple trees – we guess a Bramley and a Cox – and they not only support our family hammock, give shelter to our lawn and hide a wooden tree house, but also provide us with kilo upon kilo of apples each year.

And so was born the Hen Corner tradition of Cider Sunday.

Hen Corner is our family home in Brentford, West London – so named because we have hens in the garden and live on the corner of the street. As a business it has developed significantly over the last ten years, but it started as a hobby that simply required a brand name for our Christmas-chutney labels and social-media handle.

The decision to produce our own food was made several years before that, however, when our first child was weaning and we began to think about the food we were putting into his mouth. While I was confident of the ingredients in his baby mush, as I'd made it myself, I wondered where the carrots and parsnips had come from. Which country? Which farm? Had they been doused in chemicals? 'Fertilized' by visiting animals? How long had they been out of the ground? How many people had handled them? How many layers of packaging had they been in? How many modes of transport had brought them to us? The list went on. And then what about the cheese? What's in cheese anyway? I knew it came from milk but it didn't come out of the cow in triangles. Why were there so many different types of cheese made using the same simple ingredient?

This highchair-and-bib moment launched us on a journey that continues to this day. I started exploring the provenance of the food that we were all eating, reading the ingredients lists on packaging and questioning the impact of our food on animal welfare, our environment and the farmers who produce it – both here in the UK and overseas – as well as the benefits to our family who were eating it.

I found that the answers my questions provoked tended to be more uncomfortable than reassuring as I began to realize that most meals would include food literally flown

LEFT Cider press full of scratted apples.

in from another country. With our British dairy farms closing due to supermarket price pressures, I feared we could end up with powdered milk from distant lands as the norm. I still ask why so much of our British lamb is exported yet we're encouraged to buy meat from New Zealand – you just couldn't pick a country further away.

The information I gleaned prompted more and more questions. I just wanted to know which was the right food to buy; and as I understood more of the issues, I began to see that our food system here in the UK is pretty broken. So I had to decide: should I ignore all I'd discovered and just eat whatever I fancied whenever I fancied it (fresh strawberries with your Christmas trifle, anyone?) or should I commit to exploring more and making informed decisions? I realized that I couldn't un-know what I'd learned, and so I started sourcing the most ethical food for our family.

Organic food comes free from chemicals and with a guarantee of high animal welfare, so this was my starting point. However, it also comes with a higher price tag than less ethical products. In an attempt to offset some of that cost, and to deepen our appreciation of the hard-working farmers, we decided to try to grow some of our own food. As is the case for many others, our starting point was tomatoes, courgettes, strawberries and herbs. I thought potatoes would be fun for the children so we drilled drainage holes in an old dustbin and shovelled in compost as the plants began to grow. My aim then expanded to producing one meal for the family where the entire contents had been gathered from the garden – nothing with chemicals or plastic bags, no peas with passports, no unfair wages paid, no damage to the environment. We had fruit and veg . . . but no protein.

ABOVE Our little London kitchen.

What about a juicy steak? No room for a cow. Sausages? Where would I keep the pig? Then we saw an advert for groovy plastic chicken coops that were ideal for urban gardens and we upped our game.

We started keeping chickens in 2007 and discovered a deeper connection with where our food was coming from. They really were easy to care for and it was only months before we upgraded to a bigger coop and expanded our flock. Not long after, Andy built a cider press and we joined with neighbours to create our first home brew. We now borrow a much bigger press and in recent years have produced 73 litres (130 pints) of the stuff in an afternoon. A friend encouraged me to blog, though I wasn't sure who would want to read about our everyday lives . . . And as our following has grown we have continued to broaden the range of food that we are producing ourselves.

A key part of our journey has been, remains and I expect always will be partnering with others who walk to the same heartbeat. Fortunately, more and more people are becoming concerned about the impact of our choices on ourselves, our health, the producers and the wider environment, and there are many individuals and organizations leading the way in this food revolution. Heroes of mine include Hugh Fearnley-Whittingstall and the team at River Cottage (where I first trained in beekeeping), Jane Mason and Richard Bertinet (who taught me to bake good bread), the Women's Institute (who have championed home-produced food for generations) and Omlet (who make it so easy to keep farm animals in your London back garden).

Doing it together is so much more encouraging – you feel like you are swimming

ABOVE All hands on deck at Cider Sunday.

within a big school of fish rather than against the tide. As such, I'm proud to be a member of the British Beekeepers Association (and my local beekeeping associations), London Food Link, the Real Bread Campaign and Bread Angels. I'm happy to give my time as a director and trustee of Cultivate London and am honoured to work regularly with *Country Living* magazine, who bring country lifestyle, food and farming issues into many homes throughout the year.

As we ourselves were developing 'forgotten' country skills, we decided to see if others were interested in learning with us by offering some courses. We initially partnered with Omlet, who make the chicken houses that we use, to run our first season of courses in 2010. This began with 'Urban Hens – Keeping Chickens in London'. 'Pick and Pickle' was offered soon after, followed by 'Family, Feathers and Fun'; and by 2012, with a year of beekeeping under our belts, we offered a taster session for would-be apiarists. Running courses has been a natural progression as I've been involved in training and education for many years, both in business and in the voluntary sector; and I am delighted that Hen Corner is now drawing from the experience, skills and interests of every area of my life, all packaged up in a lifestyle that seeks to encourage, empower and equip others.

Our courses are very hands-on and, because they are run from our family home, guests can easily see how they can use their new-found skills in their own homes. I want guests to try something new and think, 'Yes, I can do this!' We've had people from all walks of life attend our courses, and for many different reasons. One woman had been given some money from her employer to try something new; her colleagues had chosen to sky dive or run a marathon and she decided to become a beekeeper with us for a day. Ironically, another woman drove from Cheddar to join in a recent cheesemaking course. And many guests of our bread course have stopped buying bread altogether because they now make it all themselves.

My favourite part of the day is collecting the eggs from our flock of twenty-seven. When I reach into the nest box, I always thank the girls and comment on the number that they've laid that day. If it's a good-sized clutch, I congratulate them – and if it's not, then I reassure them that maybe it's that time of year when the days are shorter and they're too busy growing new feathers for winter warmth, so pushing out a daily egg is just too much to ask.

"I started exploring the provenance of the food that we were all eating, reading the ingredients lists on packaging and questioning the impact of our food on animal welfare, our environment and the farmers that produce it . . ."

All the food we collect from the garden is logged into an online tool that calculates the value of each harvest. It's a real record of our reward and so encouraging to see the total increase throughout the year. In a good year, we collect over £3,000 worth of food, making us self-sufficient in eggs, honey, cider, asparagus, raspberries, courgettes and more. We then use other fruits and vegetables that we've grown to produce a wide range of jams, jellies, chutneys and pickles.

I like to cook with good food that has a good story. I choose organic where possible, high animal welfare, Fairtrade, local and so on. If food is good for me then it must be good for the wider world that its production is impacting. In terms of our growing, I have been learning as I go, trying to find low-impact ways of producing a good crop. Our commercial-sized compost bins, made from recycled plastic milk bottles, maximize the benefits of plant nutrients and chicken droppings to increase the fertility of our soil.

Back in 2015, we accidentally opened a bakery. My friend Katie and I trained as Bread Angels with Jane Mason of Virtuous Bread and decided to bake together once a month for fun, offering any spare bread to friends on Facebook. Our first Friday found us with thirty customers spending over £380 between them and saying, 'What do you mean once a month? We want this every week.' What began with friends and neighbours becoming customers soon grew as word got out, with friends of friends placing orders and a few posts on a local community Facebook group generating a lot of interest. To complete the circle, many customers have become friends over the years.

We are pleased to welcome many school and preschool groups to Hen Corner, often as part of a project exploring where their food comes from. I regularly ask children what they think the difference is between a farm and a zoo, and the replies are usually along the lines of, 'Depends what animals you want to see: lions and tigers at a zoo; sheep and chickens on a farm.' I ask why farmers keep animals and the children tell me it's because farmers are kind and like them. I've discovered that many children are completely disconnected from where their food comes from. Recently, I've been particularly focusing on how everything we eat used to be alive. Your favourite food is pizza? Well, that's flour from the plant wheat, oil from an olive tree, tomatoes, onions and herbs . . . I point to the relevant plants around the garden, then explain that cheese comes from cow's milk and the cow eats grass.

The visiting children collect that day's eggs, hug the hens and don perfectly proportioned beekeeping suits to help inspect the bees.

We've also been out to visit many schools, taking either a selection of beekeeping equipment with us or some of our very friendly chickens. Snowy, our Salmon Faverolle hen, happily sits on my lap while sixty-odd children queue up for a cuddle!

I really think that there are so many benefits and privileges of living in the city. It's wonderful to appreciate all that we have while also weaving in many country ways to enrich our lives, giving us the best of both worlds. Every choice that we make has a consequence, so let's reduce our impact on others and the environment.

With a few new skills and some good decisions, we can each turn our urban corner green.

LEFT A bakery in our own home.

MAKE

 When we've made our own food, fully appreciating the time, ingredients and, dare I say, *love* that we've put into it, we savour every mouthful and make sure that none is wasted . . .

We have found that a great way to start getting involved in our own food production is to make things ourselves. From scratch.

Understanding that everything that we eat used to be alive, whether plants or animals.

When we recognize the individual components of a recipe or a new food product, we can find a new level of respect for – and therefore value in – not only the food but also its producers, farmers and indeed the precious environment in which it has thrived.

I've recently been encouraged to use my LOAF when choosing food and I think this is a great benchmark to aim for when eating out, shopping and making our own food. This means choosing food that is:

- **L**ocal – therefore naturally in season and minimum food miles
- **O**rganic – good for us, the farmers and the environment
- **A**nimal friendly – either animal free or choosing sustainable cuts of high welfare meat
- **F**airtrade – supporting farmers and food producers wherever they are in the world.

BREAD

I remember the first time that I tasted homemade bread. (I'm not including the bread I attempted to make in school during the 1980s – because only now do I understand that it's just not possible to bake a light fluffy loaf in a school lesson, even if it was a double period.) We had been invited for lunch with friends who were then living in the house that we now call home, and I can clearly recall being delighted with the smoked-mackerel pâte served with toasted homemade bread. 'It's just made in the bread machine,' explained Ruth, our hostess, as I peered with bemusement at the gadget on the kitchen counter. 'You tip in the ingredients and press the button.' I was sold, and within hours I had researched and ordered a magical machine of my own.

I loved it, especially the overnight setting that allowed us to wake up to freshly baked bread. Then there was the setting to make rolls for the children's packed lunches; and as for the pizzas . . . amazing. I was even found taking it on holiday with us on many occasions – why have worse bread while you're away?

Fast forward several years and 'owning a bread machine' is an embarrassing confession offered up on our regular bread courses – but I'm always quick to reassure by explaining that, for many, a bread machine is a helpful step between buying a supermarket loaf packed with many unknown ingredients and baking your own artisan handmade loaf.

A typical supermarket value loaf lists ten or more ingredients, while real bread needs only four: flour, water, salt and yeast. The bread-machine recipes would ask for a few extras, but at least these were food items that we recognized from our kitchens: butter to keep the loaf softer, skimmed-milk powder and sugar to help with colour and crust. The bread machine requires these additions to compensate for the limitations of the element that bakes the bread from cold – whereas a traditional loaf is placed directly in a very hot oven, approximately 250°C/480°F.

Understanding what is in our food is definitely a good thing. Over time, though, I decided that I didn't like the bread-maker's cube-shaped loaf, with its trademark hole in the bottom, and I was using the machine only to make dough. So, when it broke, I didn't replace it.

I taught myself a basic dough, sometimes mixing by hand, sometimes in a food processor, and researched different techniques and recipes. The rest, as they say, is history.

Talking of history . . . I regularly remind myself that people have been baking bread for thousands of years – even in Genesis, the first book of the Bible, bread is mentioned

LEFT Sunflower and honey bread.

within the stories. So how did people discover or invent it? Consider the hunter-gatherers who ran after animals to roast and picked berries from bushes. These people began to settle and stay in one place when they realized that planting a single seed in the ground could result in multiple seeds being grown if they only waited around long enough. After they'd harvested many seeds, they discovered that they needed to crack open the grains from the ancestors of our modern wheat to reach the nutrition inside; otherwise the outer husk would preserve the grain, allowing it to pass through their bodies whole and bringing them no benefit. So, they began to grind their harvested grains between stones to unlock the goodness hidden within. And here we have the first flour.

Flour is the main ingredient of bread and gives us energy, nutrition and sustenance. But raw flour is neither easy nor pleasant to eat, so adding water to make a paste or porridge helped to make it more palatable. The transition from porridge to loaf may well have been an accident. Yeast is everywhere and without moisture or a food source will lie dormant. However, when the natural yeasts on these ancestral grains were woken with a splash of water they would have found all the lovely carbohydrates (now easily accessible, thanks to the stone-crushing) to feast on. As I explain to schoolchildren, while the yeasts enjoyed their banquet they'd also burp out bubbles of carbon dioxide in their overindulgent excitement.

The first loaf of bread may well have been a bowl of wheat porridge left overnight and – as food was never wasted due to the hard work needed to produce it – warmed by the fire or maybe dropped on a hot stone. The yeasts would have continued to ferment until the cooking temperature killed them, leaving the hungry hunter-gatherer souls with a warm, slightly fluffy flatbread – much nicer to eat than the sticky porridge of yesterday.

Today's bread is still based on the principles of mixing flour and water and the presence of some yeast, be that commercially produced or wild yeasts cultivated into a sourdough starter. The only other ingredient we add now is a teaspoon of salt, which not only adds flavour but also brings a bit of stability to the fermentation process.

So, the first breads were what we would now call sourdoughs, fermented using a flour-and-water paste that nurtures wild yeasts and is regularly fed more flour to keep it active. In later years, leftover yeast from beer-making would be collected from the local brewer to make a sweeter-tasting loaf; and bakers all over the world have been experimenting with flavours and techniques ever since the discovery that good bread fuelled you with energy, filled you with nutrition and tasted great.

You may be familiar with the saying 'the best thing since sliced bread'. While I appreciate the convenience that it brings, the invention of the supermarket value sliced white loaf wrapped in polythene has unfortunately contributed towards our collective dependence on cheap food and devalues the necessity of good ingredients, skill and time required to produce high-quality food that is better for us, better for the environment and better for the producers.

I completely appreciate that postwar ration-dependent Britain was striving to find an affordable way to feed the masses amid challenging times, but the highly 'efficient' method known as the Chorleywood bread process ended up causing more problems than it solved. It relied on big machinery turning day and night, adding unrecognizable ingredients and numerous additives to rush through as many loaves as possible. This process depended on so many shortcuts it's no surprise that many people today believe they have an intolerance to bread – whereas it may just be that they are unable to digest the adulterated loaves.

While developing my own bread-making skills, I was following the adventures of Jane Mason of Virtuous Bread, who had started the Bread Angels network. Here, people around the world make real bread available to their communities both as loaves to buy and through teaching skills that equip others to bake their own beautiful bread. When the opportunity arose to train as a Bread Angel myself, I signed up to increase my experience and advertise the bread courses I had begun to teach.

As I trained with my friend Katie, we had no intention at all to bake bread for sale, but the popularity of our first endeavours was such that our accidental bakery came to life. Within six weeks we had set up an online store to manage the orders and here we are now many years later.

There are three main stages when making bread, with your dough left to rest allowing it to ferment (or prove) in between. The whole process, from beginning to end, will take around three hours, but only thirty minutes of that is hands-on time.

STAGES OF BREAD MAKING

Mix the ingredients well, then work the dough, stretching and folding it over itself as though you're coiling a spring. This develops the gluten needed to form stretchy 'balloons' within the dough to hold the bubbles of carbon dioxide, and also creates tension for even proving. Rest the dough while it ferments for around an hour.

Shape the dough into your chosen loaf – whether that's rolls, a plait, a baguette or for a tin. Shape the dough tightly, folding all seams and edges into the middle of the dough and leaving a smooth, taut surface on top. Rest the dough while it ferments for around an hour.

Bake the dough in a very hot oven until it's brown all the way round and sounds hollow when you tap it. Allow to cool completely before cutting: this avoids breaking the crust, which protects the loaf as it finishes its transformation from gathered grain to magnificent meal.

BEAUTIFUL BREAD

Give us today our daily bread . . .
Be it Indian naan, Italian ciabatta, French
baguette, Irish soda, Eastern flatbread,
Greek pitta or Russian black bread,
billions of us around the world are eating
bread every day. This recipe is a great
foundation that would happily welcome
seeds, nuts, fruit and experiments with
different flours.

MAKES 1 LOAF OR 8 ROLLS

INGREDIENTS

500g/1lb strong
    white (wheat) flour
1 tsp dried yeast
1 tsp salt
350ml/12fl oz water

METHOD

1. Measure the dry ingredients into a large mixing bowl,
   followed by the water.
2. Draw the ingredients together with your hand to combine
   well and squeeze the dough into a ball.
3. Work the dough for 15 minutes, initially in the bowl and
   then on a clean work surface, pulling and stretching it
   over itself.
4. Pull the dough into a ball and place in a large clean bowl,
   cover to stop it drying out and leave in a warm place to
   prove for 1 hour.
5. Preheat the oven as high as it goes (220°C+/430°F+).
6. Fold the air out of the dough, adding additional
   ingredients if you wish, and shape as desired. Place in a
   tin or on a lined baking sheet, cover and leave to prove for
   up to 1 hour.
7. Bake for 20 minutes, or less if you've made smaller rolls.
8. Check that your loaf sounds hollow and bake for a bit
   longer if not. Cool on a wire rack, then enjoy.

CINNAMON SWIRLS

These are the top sellers at our micro bakery – every single week. We use an enriched dough for this recipe, adding eggs, butter and sugar to a traditional basic bread dough. This dough can also be used for iced fingers, hot cross buns and many more sweet treats.

MAKES 12

INGREDIENTS

For the dough

500g/1lb strong white (wheat) flour

1 tsp dried yeast

1 tsp salt

40g/1½oz caster sugar

250ml/9fl oz milk

60g/2oz butter, room temperature

2 eggs

For the filling

50g/1¾oz soft butter

50g/1¾oz soft brown sugar

2 tsp ground cinnamon

METHOD

1. Measure the dry ingredients into a large mixing bowl, followed by the butter, milk and eggs.

2. Draw the ingredients together with your hand to combine well and squeeze the dough into a ball.

3. Work the dough for 15 minutes, initially in the bowl, then on the clean work surface, pulling and stretching it over itself.

4. Pull the dough into a ball and place in a large clean bowl, cover to stop it drying out and leave in a warm place to prove for 1 hour.

5. Fold the air out of the dough and roll it out into a rectangle 1cm/½ inch thick.

6. Spread the soft butter over the dough, then sprinkle with soft brown sugar and ground cinnamon.

7. Roll the rectangle up into a baton. Slice the baton into 12 rounds and place them cut side up on a baking sheet lined with parchment. Cover to stop them from drying out and allow to prove for 1 hour.

8. Bake at 180°C/350°F for approximately 12–16 minutes. Look for firm, risen, golden swirls.

9. You can glaze or ice these if you choose or simply dust with icing sugar.

MARMALADE

I t must have been a magazine recipe that first encouraged me to make marmalade. I don't remember why exactly; I just recall that I started it one evening, many years ago, after putting the kids to bed. And I made a lot of mess. And it took ages. As we now generally make marmalade in January, to coincide with the Seville orange harvest, it might have been a New Year, New Project thing, or maybe I was a bit bored after all the busyness of Christmas, or simply a case of 'I want to do that' – a trait of mine that has led to many new adventures over the years.

I checked the ingredients, bought four times what was needed – well, if I was going to make some, I may as well make lots! – and set aside an evening to get productive. The first part of the process was to boil the oranges along with a couple of lemons: no problem, but it did take a couple of hours. Then I had to chop them up into tiny slithers, which wouldn't have been so bad if I hadn't chosen the 'four batches at a time' route. It was therefore around 11 p.m. when I was ready to start making the actual marmalade. In my usual enthusiastic way, I took my biggest saucepan and filled it to the top with chopped oranges and sugar – confident that this wouldn't take long and I'd soon have multiple jars to stash away, self-sufficient in golden yumminess.

Unfortunately, I didn't understand the process, or what was supposed to happen.

The recipe told me to bring the fruit and sugar mix to a fast boil, yet just as it started to bubble it began to overflow. *This doesn't happen to water when it boils so why is my magic marmalade expanding so much?* I turned the temperature down and the pan contents receded back into place – phew! But now it wasn't boiling, so I turned the heat back up and it rose again and began to drip down the outside of the pan, catching with a sizzle on the hob's gas flame. *Drat, it's doing it again.* After a few attempts to bring it to the right temperature without it boiling over, I realized that I had just way too much mixture for the pan – four times too much to be precise . . .

I found some other saucepans, which weren't quite as big, and poured out half of the 'citrus lava'. This still left too much in the pan. I rinsed out our wok (to remove any trace of soy sauce) to add to my armoury of vessels and even ladled a jar's worth into our frying pan to redistribute the mixture. About an hour or so later, I had a pot or pan on each of our hobs and was carefully watching to make sure none of them boiled over, and eventually I saw the high foaming boil synonymous with jam-making. The fruity sugar needs space to quadruple in size, as excess water evaporates and the boiling process releases pectin: the magic ingredient that turns fruit juice into jelly.

LEFT Homemade marmalade on homemade toast, yum!

Fortunately, I ticked this off as a steep learning curve and have been making marmalade every year since. Some of it has even won Gold in the World Marmalade Awards.

A BRIEF HISTORY OF MARMALADE

We've been making marmalade in England since the seventeenth century, based on recipes passed down from Roman times. Before the production and transportation of cheap sugar, honey was the sweetener of choice and this is what precious fruits such as quinces would be poached and preserved in, allowing them to be enjoyed throughout the winter months when fresh fruit was scarce.

Now that we have access to both Fairtrade cane sugar and locally produced sugar from beet, we can transform all kinds of fruits into long-life delicacies. It seems perfectly natural to preserve a favourite fruit, to suspend it in time as you suspend it in sugar – but why marmalade; why oranges? I may eat oranges by the segment, or drink their juice, but I'd never eat a whole orange, skin and all – so what makes the citrus family perfect for the job?

I think that the secret is in the skin of the citrus fruits (the hallmark of marmalade): here are found the essential oils renowned for decreasing stress and generally lifting your spirits. What better way to start your day in the middle of winter?

The firm, bitter oranges of Seville are believed to make the best marmalade in the world due to their sharp flavour, which complements the sugar, and their high pectin content for the perfect set. But don't try to buy them while you're in Spain, as they are all shipped over to the UK.

LEFT Boiled oranges and lemons, ready for slicing. **RIGHT** Seville oranges shipped from Spain.

SEVILLE ORANGE MARMALADE

Don't try to buy
Seville oranges in
Spain, they are all
shipped to the UK.

MAKES APPROXIMATELY
10 X 340G/12OZ JARS

INGREDIENTS

1kg/2lb Seville oranges

2 lemons

2kg/4lb granulated sugar

METHOD

1. Boil the oranges and lemons (whole) in a large preserving pan with 2.5l /4½pt of water and simmer very gently for around 2 hours, ensuring they don't boil dry, until the peel can be easily pierced with a fork.

2. Saving the cooking water, lift out the fruit to cool, then cut each one in half. Scoop out all the pips and pith and add to the reserved orange liquid in the pan.

3. Bring to the boil for 6 minutes, then strain this liquid through a sieve into a bowl and press the pulp through with a wooden spoon – it is high in pectin so will give your marmalade a good set.

4. Cut the peel, with a sharp knife, into fine shreds. Add, along with the cooking liquid, to the preserving pan with the sugar.

5. Stir over a low heat until all the sugar has dissolved, for about 10 minutes, then bring to the boil and bubble rapidly for 15–25 minutes until setting point is reached, 105°C/221°F.

6. Leave the marmalade to stand in the pan for 10 minutes to cool a little and allow the peel to settle. Pot in sterilized jars (see instructions for sterilizing on page 92), then seal and label.

HONEY, LEMON & GINGER MARMALADE

While we all love a traditional orange marmalade, we've created this new recipe using luscious lemons and honey from our bees, with some cheeky chunks of stem ginger, to brighten up your breakfast and make your supper snacks truly special.

MAKES APPROXIMATELY 10 X 340G/12OZ JARS

INGREDIENTS

1kg/2lb lemons

250g/8oz honey

1.75kg/3½lb granulated sugar

6 pieces of stem ginger in syrup, chopped

METHOD

1. Boil the lemons (whole) in a large preserving pan with 2.5l /4½ pints of water and simmer very gently for around 2 hours, ensuring they don't boil dry, until the peel can be easily pierced with a fork.

2. Saving the cooking water, lift out the fruit to cool, then cut each one in half. Scoop out all the pips and pith and add to the reserved lemon liquid in the pan.

3. Bring to the boil for 6 minutes, then strain this liquid through a sieve into a bowl and press the pulp through with a wooden spoon – it is high in pectin so will give your marmalade a good set.

4. Cut the peel, with a sharp knife, into fine shreds. Add, along with the cooking liquid, to the preserving pan with the sugar, honey and ginger.

5. Stir over a low heat until all the sugar has dissolved, for about 10 minutes, then bring to the boil and bubble rapidly for 15–25 minutes until setting point is reached, 105°C/221°F.

6. Leave the marmalade to stand in the pan for 10 minutes to cool a little and allow the peel to settle. Pot in sterilized jars (see instructions for sterilizing on page 92), then seal and label.

PASTA

Wheat is wonderful isn't it? Daily bread, cakes, biscuits, pastry, scones . . . the list is endless. It's not difficult to look back and discover that we've eaten wheat in some format for every meal of the day. But leaving baked goods aside, I'd love to shake the hand of the person who first discovered that combining flour with egg, and working it into a smooth, firm dough, could create a product that could be rolled, folded, stretched, cut, stuffed, boiled and baked and would become a family staple.

Mamma mia . . . Be it carbonara, lasagne, vongole or pesto, a perfect pasta is the staple base for many of our best loved dishes.

I've enjoyed many meals in my time, but my absolute favourite is spaghetti Bolognese. And I don't want to get into the discussion of how the people of Bologna would never put their meat sauce with spaghetti: apparently they prefer the ribbons of tagliatelle. I love mine made with minced beef, mushrooms, onions, garlic, tomatoes and a good splash of red wine. For me, long slender strings of spaghetti are perfect, and there's no shame in sucking up the unruly cords that trail from the fork, splattering tomato drips in the process.

But before we reach for a packet of dried pasta, in whatever shape, size or colour that you fancy, have we ever thought that it might not be too difficult to make from ingredients that we already have in the kitchen and, if you keep chickens, maybe from the garden too?

When our son James was in primary school, his teacher asked if I would like to do a cooking lesson with the class. Of course, I jumped at the opportunity. I asked what facilities they had and learned it was a very basic electric hob with a mini oven on a trolley that could be wheeled from room to room. So I started planning. *Let's try and make a meal including lots of food that we gather from the garden. We've got eggs from the hens, tomatoes, onions and garlic. Great, all I need from the kitchen is a bag of flour, olive oil and salt and pepper.*

As I began the lesson, I asked, 'Who likes pasta in tomato sauce?' and all hands shot up in an affirmative response. This was a good start. 'Has anyone ever made pasta before? If so, what do you do?' This was met with a few initial suggestions such as, 'Open the tin?' and as I probed for more – this was a cooking class, after all – a little girl beamed with confidence as she proclaimed: 'I've seen my mum make pasta – you put hot water in a saucepan and open the packet . . . '

OK, we were starting from scratch, but that was the point of the lesson – and it was a great opportunity to go through every ingredient, explaining how it grows and its contribution to the dish.

LEFT Fingers in the flour.

We weighed out the flour and took it in turns to squash and stretch the dough; the hand-cranked pasta machine was a hit and it wasn't long before we had long sheets of silky pasta ready to be sliced, as if by magic, as we turned the handle the final time.

Meanwhile, in the little oven, we were roasting the home-grown veg in a little oil seasoned with salt and pepper – this was going to be great.

We boiled water in a large pan and carefully lowered down our freshly cut ribbons into the bubbles below. Fresh pasta only takes a couple of minutes to cook, and our finished dish was ready to sample in no time.

Unfortunately, the response was mixed. I was encouraged by those enthusiastic to taste the creation that they had contributed towards – 'Best meal ever!' – but sadly there were some children who flatly refused to taste it. Even though they had been involved in the process of making it and recognized all the components of the dish, it was unfamiliar and not like the food they ate at home. I'm fortunate that my own children have always eaten everything they've been given and have a healthy appetite to try more unusual foods. It's a shame when children don't want to try new things, as a limited diet can impact health, become inconvenient to accommodate and can go on to become a challenge in social occasions such as eating out, staying with friends and going on holiday.

Once you've tasted your own handmade pasta, even served simply with butter or a drizzle of olive oil, maybe with a grating of cheese on top, you'll see it's worth the prep time for something so delicious.

We make an egg pasta dough that can be rolled, sliced, folded or wrapped around your favourite filling for an authentic homemade supper.

LEFT Simple ingredients for pasta dough. **RIGHT** The whole family can get involved.

EGG TAGLIATELLE

You'll need a hand-turned pasta machine for this recipe. You can easily adapt the recipe for lasagne, and for filled pasta just take it to the thinnest setting on the machine at step 6.

SERVES 4

INGREDIENTS

400g/13oz Italian 00 white flour or strong bread flour

4 eggs

METHOD

1. Measure the flour into a large mixing bowl, make a well in the middle and crack in the eggs.
2. Starting with a fork, beat the eggs together, slowly drawing in some flour, then use a scraper to combine the eggs and flour into a ball.
3. Knead the dough for 15 minutes. Pull into a ball, place in a plastic food bag and leave to rest for 30 minutes.
4. Using half the dough at a time, start feeding through the pasta machine on its widest setting, folding in thirds, like a letter, in between rolls and rotating 90 degrees. Have flour for dusting to hand in case the dough becomes sticky.
5. After 4–6 rolls, start to reduce the thickness of the pasta sheets on the machine settings. Repeat step 4, working your way down through the settings on the machine and stopping at the second-thinnest.
6. Finally, pass the pasta sheets through the cutting rolls and gather the ribbons below.
7. If you're not cooking the pasta straight away, dust it in flour and semolina to assist drying and then hang it in lengths.
8. Cook in boiling salted water for 2–3 minutes, or slightly longer if it's been dried.

SQUASH & SAGE RAVIOLI

Fill me up! Be it a square ravioli or a hand-shaped tortellini, a delicious filling makes all the difference. Be creative with your favourite flavours, using meat, cheese or other treats for a firm paste to stuff your pasta. Here is one of our favourites . . .

SERVES 4

INGREDIENTS

1 batch of fresh pasta in sheets (see page 32)

½ a butternut squash

1 onion, chopped

2 cloves of garlic, chopped

1 tbsp olive oil

125g/4oz ricotta cheese

1 handful of sage leaves, torn

nutmeg, cinnamon, salt and pepper to taste

METHOD

1. Peel and deseed the butternut squash, cut into chunks and place in a roasting tin with the onion, garlic and olive oil. Tuck the torn sage leaves in between the vegetables. Season with salt and pepper and roast for 30–40 minutes at 180°C/350°F.

2. When the squash is soft and starting to caramelize at the edges, remove from the oven and transfer, with the onions, garlic and any juices, to a clean bowl.

3. Add the ricotta cheese, mash together and season to taste with nutmeg, cinnamon, salt and pepper.

4. Lay out a sheet of fresh pasta and evenly distribute small spoonfuls of filling on top, then brush water between the mounds of filling. Lower another sheet of pasta on top and press down around the filling to seal, trying to exclude air from the tiny parcels.

5. Cut through the pasta sheets to create individual pieces of ravioli.

6. Cook in boiling salted water for 3–4 minutes.

SAUSAGES

I love sausages.

I love them barbecued, in a casserole, in a sandwich and with mash. I love pigs in blankets, Scotch eggs and toad-in-the-hole. I *LOVE* sausage rolls, especially my homemade ones.

Yet the way that I cook sausages most often is squeezing the meat out of the skins and frying it as a base for a pasta ragù sauce.

In our house, as we love all food that comes from a pig, we call pork products 'Vitamin P'. However, when you look back in time and around the world, you'll find that sausages aren't necessarily pork (as the majority of British bangers are) – sausages were and are made with what's available.

Sausages are so versatile and their creation – over four thousand years ago, according to a stone tablet found in Mesopotamia – provided the perfect way to reduce food waste. They were originally devised as a handy way to use up trimmings and offal, packing it all together with some good seasoning into a cleaned-out tube of intestine from a nearby just-slaughtered animal.

Now, if animals are being kept for meat, their welfare during life and death are paramount; and, if that animal has died for our dinner, it's not only thrifty but also honouring to eat the whole animal – out of respect for its life. As with all food, I want to make sure that nothing is wasted. And pigs are well known as the animal where 'you can eat everything but the oink'.

Today's factory farms and highly processed foods seem a million miles away from keeping a pig in the backyard and processing it on the kitchen table. But, just like the connections between farmers in the countryside, those of us in towns and cities can benefit from communities that share our passion for producing our own food and support us on the journey. We can find other like-minded people who are sharing their gardens and open spaces with animals for food. There are opportunities to volunteer, form collectives and get hands-on when it comes to sourcing animal-friendly meat.

While I usually buy our pork directly from the farmer, often half a pig at a time, we shared a pig once (for more about this, see page 137 in 'Keep') and I was very much looking forward to processing as much of it as possible. So, when it was sent to the abattoir before its return via the butcher, I was asked, 'Do you want the liver and kidneys?' *Yes.* 'The trotters?' *Yes.* 'The head?' *Yes, if no one else wants it.* I scoured books

LEFT A family favourite – sausage rolls.

to prepare me; braised trotters, brawn from the head, bacon of course . . . now, could I even make salami? (I did.)

I went on a sausage course, which was not only helpful but also great fun, and it was there that I understood the role of salt in both seasoning and preserving: our modern-day word 'sausages' originates from the Latin word *salsus*, meaning 'salted'. I also learned to recognize that breadcrumbs in sausages aren't a bad thing, because they help absorb the fats and juices that carry the flavours. On the course, we followed a couple of tried-and-tested recipes, but now I like to experiment with favourite flavours and seasoning combinations.

When I make sausages at home, pushing the meat through the mincer, I often find myself singing a line from the musical *Les Misérables*: 'Kidney of a horse, liver of a cat*/ Filling up the sausages with this and that . . . ' And, yes, sausages can be made with almost anything – but when you get the perfect amount of this and that, you can create something truly wonderful.

*Please note that I'm not recommending including your pet cat in the recipe.

My favourite flavours are:

- Red wine, chilli and fennel seed
- Apple and sage
- Chutney (add a spoonful into the mix)
- Marmalade (add a spoonful into the mix)

We test the mixture by cooking a teaspoon at a time, just to check that we've got the seasoning right; then, once we've got it spot-on, some goes into sausage skins, some is wrapped in puff pastry for sausage rolls, some is moulded around hard-boiled eggs before being rolled in breadcrumbs and deep fried as Scotch eggs, and some is simply bagged up to stuff a chicken or make a sauce for pasta. As long as you've made the sausages with fresh meat, the raw sausage can be frozen until you want to cook it.

On a recent course that I was running, a guest said, 'I need to know how to make sausages for my children without using a sausage machine as I've no space to store one. ' No worries: we were eating sausages in this country long before the electric machines were invented. I handed her a large funnel and she was very successful in her stuffing.

Select your own seasonings and stuff away . . .

ABOVE LEFT Pass the meat through a mincer.
ABOVE Sausagemeat ready for stuffing into skins.

SAUSAGES

Get creative and make up your own flavours.

MAKES APPROXIMATELY 8

INGREDIENTS

1m/3ft 3in length of natural sausage casings

250g/9oz pork shoulder, diced

250g/9oz pork belly, diced

50g/1¾oz breadcrumbs (10 per cent of the meat's weight)

1 tsp salt

freshly ground pepper

your chosen flavours, spices and seasonings: fennel seeds, chilli flakes, dried or fresh herbs, garlic, nuts, wine, cheese, chutney, etc.

METHOD

1. Soak the sausage casings in cold water overnight. Rinse inside and out with warm water when you are ready to begin.
2. Pass the pork shoulder and belly through the mincer, collecting in a clean bowl.
3. Add breadcrumbs and salt to the minced pork along with your chosen flavourings, then combine well using your clean hand. Start with smaller quantities of seasoning; you can add more after tasting.
4. Fry a teaspoon of the mix to check for flavour, then adjust seasoning if desired.
5. Attach sausage filler and casings to the mincer and proceed to fill a long length of casing. This can be done with a piping bag or large funnel if you don't have a mincer/sausage machine.
6. Twist the filled casings into sausages and rest overnight in the fridge.
7. Cook from cold in a cold pan/grill/oven to reduce splitting and never prick your sausage or you'll lose your lovely flavours.

CHEESE

'I know cheese comes from a cow but it doesn't come out in triangles . . . ' I love asking schoolchildren where cheese comes from, to watch their faces screw up trying to guess how milk, which they are familiar with, can be transformed into their favourite pizza topping. Answers have ranged from 'You put something into the milk to make it go hard,' to 'Maybe you freeze it to turn it into a solid . . . ' I try to be gentle when asking them to think how a frozen block of milk could be grated into a sandwich and the giggles, while imagining soggy bread, show that I haven't teased them too much.

I then go on to explain that milk can become a solid when we take the liquid out – this seems to cause more confusion, as milk is already a liquid. 'Maybe you sieve milk to make cheese?' comes another hopeful suggestion. The answer can be found in the old nursery rhyme about Little Miss Muffet, who sat on her tuffet eating her curds and whey. If we add something to fresh, warmed milk, usually rennet (originally from a sheep's stomach, although many vegetarian versions are now available; and in the case of paneer cheese it's lemon juice or vinegar), we can cause the milk to coagulate. This is like clotting, where the solids find each other as curds and the remaining liquid, which can be drained off, is whey.

I remember my mum draining milk that was already past its best through a cloth in the 1970s, telling us this was how cottage cheese was made, but my first real understanding of the process came when we took our children to Lynher Dairies in Cornwall, where Yarg cheese is made, when they were very young. We were able to meet the cows, stroking their noses before they were walked through to the milking parlour. We watched them get hooked up to the pumping machines and it wasn't long before gallons of milk swooshed through clear pipes away to the dairy. We were then taken to watch the cheesemaking from a viewing gallery and saw huge tanks filled with milk. We were very surprised to see someone drag what looked like a huge comb through the milk and cut it into slices. I later discovered that this milk had already been warmed and had had the rennet added. The curds were then stirred to help them break up into small pieces. Moving along the viewing gallery, we saw curds being packed into round moulds to support them as the remaining whey drained off and then, finally, the blocks of cheese being wrapped in nettle leaves. The process was fascinating and the finished cheese that we tasted at the end was delicious.

Britain produces a huge range of cheeses – I believe we make even more varieties that you can find in France. I'm sure our climate must have something to do with it. Cows need grass, grass needs rain and we get lots of rain over here.

LEFT Feta cheese with pomegranate seeds.

Another family holiday took us to Billingsmoor Farm in Dorset, where we were able to spend time with the farmers, Jayne and Robert Lammie, and learn more about organic dairy farming. Unfortunately, dairy farmers in the UK have struggled considerably over the last twenty years or so. With rising costs and supermarkets wanting cheap milk, many have found it completely unsustainable to continue and have sadly given up milk production. But Billingsmoor Farm is among those whose owners have persevered, farming organically and diversifying, and there's now a fabulous campsite there. Hopefully the cows can continue their twice daily visits to the milking parlour.

I first started making cheese at home when my good friend Rachelle came to visit from Australia and told me how her neighbour, who owned a cow, had turned up on the doorstep with a bucket of milk one day, asking if she could use it. Just like me, Rachelle couldn't bear to see food wasted – but, after finding no space in her fridge, she'd realized that she needed to think of another way to utilize this fabulous gift.

Of course, cheesemaking is the perfect way to preserve milk. Way back before refrigeration, a jug of milk could be kept in a cool place for just a few hours, but beyond that it would go sour and lumpy very quickly. While the first cheese was probably an accident, the very nature of removing moisture and including salt reduces bacterial growth and helps keep food safe to eat.

SO HOW DID CHEESE GET MADE BY ACCIDENT?

We know that people have been making cheese for thousands of years. I found a little line in the Bible story of David and Goliath where the young shepherd boy wanted to visit his older brothers on the battle line as they fought the Philistine army (who had the giant Goliath as their hero). David's mum said, 'No! It's enough that your older brothers are away; we need you here to help us – those sheep won't look after themselves,' or words to that effect. But his dad, Jesse, said, 'Take these ten loaves of bread and ten cheeses to your brothers, see how they are doing and bring back good news.'

This got me thinking. Imagine the young David, going out for long days in the heat, walking his flock of sheep along the mountainside looking for fresh patches of grass that they could chew on. He would have to take supplies for the day and sheep's milk – being in plentiful supply and a source of both nutrition and hydration, food and drink – would have been the perfect packed lunch for the growing lad. Obviously, these were the days before plastic bottles, but never fear: a popular way to store liquid was to use a skin – in particular, the stomach from a dead animal, such as a sheep. You would clean it, obviously, and then tie a knot in each end to hold the liquid secure.

David's mum had both milk and a sheep-gut bottle, so off he went each morning to work. Maybe David fell asleep in a cave and left his milk by accident? Maybe he saw some tempting berries or a small bird that he could catch for a hillside BBQ? Whichever, when he returned to his milk later in the day, it had set solid. A few shakes would break up the curds and on tasting he perhaps found it quite pleasant, albeit quite different from the drink he was expecting. A sprinkling of salt would add to the flavour, help it to firm up and keep it fresh for longer.

LEARNING FROM RACHELLE

Rachelle's first experiments with the bucket of milk from her neighbour were inspired by Internet research and utilized equipment and other ingredients already at home, but she soon started exploring different recipes for cheese and discovered that, while milk from different animals will affect the taste of the finished cheese, it's pretty much all curds and whey – with the key differences between varieties being time, temperature and technique.

During her and her family's stay with us, Rachelle bought moulds and cultures to teach me how to make Camembert (did you know that white rind is actually mould that you grow around the cheese?) and we bought some clean rubber gloves to make mozzarella, which involves warming up the curds in hot whey and stretching them before rolling them up in a ball.

After her visit, I started playing around with other recipes and techniques. I have made a lovely firm blue cheese, where the milk is inoculated with *Penicillium roqueforti* and, after piercing the firm cheese with sterile needles to allow oxygen to reach the centre of the block, beautiful blue veins start to spread throughout.

Unfortunately, most published cheese recipes, and additional ingredients such as cultures, are based on batches using 1,000 litres/1,760 pints of milk. That's not going to work in my kitchen! So, I've been adapting recipes that work really well for home use in a domestic kitchen. Not only have my family loved my cheeses but I've also been asked to share my recipes with a local Women's Institute group and am working regularly with a high school teaching cheesemaking on their Food and Nutrition GCSE course.

To make your own cheese, you need to start with some good milk. I tried making cheese with goat's milk from the supermarket but it was so highly processed I couldn't separate it.

ABOVE Hen Corner blue cheese.

KEY WORDS ON YOUR MILK BOTTLES AND WHAT THEY MEAN

ORGANIC

The cows have a high-welfare life, predominantly outside eating grass. All feed is organic and there's no use of routine antibiotics. So the milk is as natural as can be.

WHOLE

The cream within the milk has a higher fat content so would naturally float to the top of the tank. For a lower-fat milk, this cream is skimmed off or partially/semi-skimmed. Whole milk is full fat.

PASTEURIZED

This process, developed by Louis Pasteur, involves heating milk to 71.7°C/161.06°F for 15–25 seconds before cooling it down quickly. This kills bacteria, making the milk safe to drink.

UNHOMOGENIZED

Homogenization is the process of blitzing up milk to reduce the size of the fat particles to help with even distribution. While this stops the cream floating to the top, I find that it's easier to make cheese with unhomogenized milk that hasn't been through this process.

MAKING CHEESE AT HOME

While there are some basic similarities within all cheese recipes – separating curds and whey, and using salt for flavour and preserving – a few extra ingredients will help give consistently good results.

- **Mesophilic starter** – This helps the good bacteria in milk flourish and assists with curd formation.
- **Rennet** – This was originally extracted from the lining of a sheep's stomach, but vegetarian versions have now been developed using nettles, bark and other plants that have the necessary coagulating properties.
- **Citric acid** – Helps the milk separate for cheeses like mozzarella.
- *Penicillium candidum* – A culture for white-mould ripened cheese such as Brie and Camembert.
- *Penicillium roqueforti* – A culture for blue-veined cheeses such as Stilton and Gorgonzola.

While some recipes encourage whey to drain from curds in moulds or baskets, others extract more whey through using higher temperatures. Paneer is wrapped in a cloth and the whey is squeezed out; and cheddar has salt sprinkled on the chopped curds to extract more moisture before transferring it to a cheese press for a nice firm structure.

If you haven't made cheese before, have a go at this easy recipe for a fabulous fresh cheese that is ready to eat the next day and is so much cheaper than buying it in the shops.

RIGHT Separating curds from whey.

FETA CHEESE

Firm and tangy, wonderful in salads and quiches.
This cheese is easy to make and stores well in
the fridge – the cheese of summer.

MAKES APPROXIMATELY 500G/1LB

INGREDIENTS

4 litres/8 pints organic whole milk
rennet – use quantities stated on packet

For the brine
1 litre/1.8 pints boiling water mixed with 50g/1¾oz salt

METHOD

1. Dissolve the rennet in half a cup of cool boiled water.
2. Gently warm the milk to 30°C/86°F, then add the diluted rennet and gently stir, up and down, for several minutes. Cover and leave to set for 1 hour.
3. The milk will have coagulated to a solid gel-like form. Cut the curds into 2cm/¾ inch cubes using a long knife and cutting from different angles.
4. Gently stir the curds for 10 minutes to release more whey from them.
5. Pour the curds into a colander lined with cheesecloth or a clean tea towel, collecting the whey in a bowl underneath.
6. Tie the corners of the cloth so you can hang it like a bag from a hook overnight – we use a cupboard door handle for this.
7. Unwrap and store in cool brine in the fridge.

GROW

 Now that we're making some of
our own food, it's time to sow
seeds for the freshest ingredients.

If we really want to understand and appreciate where our food
comes from, there's no better way than growing our own. Be that
cheering on a tiny aubergine, praying that it will grow big enough to
eat before the frost hits, or twisting off yet another courgette
knowing that it will be equally fabulous in
a cheesy quiche, chocolate cake or spicy chutney.

Everyone can have a go at growing their own food, maybe
starting small with some herbs, chillies or microgreens on a
windowsill. If you have some outdoor space, a pot of tomato plants
can yield a good crop; and being able to allocate space for a veg
patch or fruit trees can open up the possibilities even wider.

Once you've tasted homegrown asparagus cooked within minutes
of cutting the spears, you won't be interested in the supermarket
bundles wrapped in plastic anymore. Humble potatoes scrabbled
from the ground, quickly boiled and slathered in butter is a dish fit
for royalty. And when your fruit tree offers more plums than you
could possibly eat and you have to choose between giving them
away or making jam for winter . . . well, both options sound pretty
good to me.

STARTING SMALL

When schoolchildren come to visit, exploring 'Where does my food come from?' as well as watching the bees, guessing which tree will bear what fruit and hugging hens, of course, we also take time to sow some seeds to be taken home. Not only is this a good lesson in horticulture – learning how to care for a newly sown crop, waiting patiently for germination and then the big responsibility of ensuring that this baby plant gets the right balance of sunlight and water – but in that process of sowing a seed that will produce food to eat, we step into a new role, akin to that of a farmer. Now, please don't think that I am comparing growing herbs on the kitchen windowsill to the huge responsibility that our farmers carry out every day of every year, feeding our nation come rain or shine. But any understanding and valuing of the hard work our farmers do, often under extreme pressure, is a step in the right direction.

If there's one lesson I've learned over the years, it's that growing your own food isn't easy. However, starting small, with something that you'll enjoy eating, is a great way to begin and is so rewarding.

Choosing a city lifestyle often means sacrificing outdoor space of our own, but fortunately there are many crops that prefer a sunny windowsill to a large field – chillies, for example, and tomatoes thrive when grown in pots or growbags. As for a balcony, many metres up,

WHAT DO PLANTS NEED?

When we understand what plants need and why, then we can plan and care for them in the best way.

WATER	NUTRIENTS	SUNLIGHT	SPACE	SUPPORT	POLLINATION
to hydrate the seed and start germination; also to transport nutrients around the growing plant.	in the soil or compost, to feed the growing plant.	which the plants turn into food by a process called photosynthesis.	to grow bigger. This means enough space in terms of pot size for the roots and above for the plant, leaves and fruit.	for climbing plants.	if we want the plant to produce fruit.

LEFT Tomato seedlings in fibre pots.

SPROUTING BEANS

This must be the quickest way to grow food indoors and you don't even need any compost. Mung beans are perfect for Chinese-style beansprouts and sprouting chickpeas are great in a salad.

Rinse half a cup of dried beans in fresh water before soaking overnight in a large container of water; watch out as they swell to double the size.

Drain the beans, rinse again, transfer to a colander over a bowl for airflow, and cover with a tea towel, mimicking the dark environment under the soil.

While they are germinating/sprouting, rinse with fresh water a couple of times a day.
Mung beans will sprout within a couple of days and can be used in salads or stir fries.
Chickpeas will need a longer time to germinate and sprout, as they are a bit bigger.

your carrots will be protected from the pesky carrot fly that hovers no higher than 60cm/24 inches and happily lays her eggs around the shoulders of young carrots in the ground.

HARVEST IN YOUR HOME

If you have no outside space at all, you can choose crops that grow well in pots. It's a good idea to start seeds off in smaller pots. I use biodegradable fibre pots, which are perfect for germination and can fit neatly into a seed tray that can be slid inside a clear plastic bag for germination to keep the seeds warm and stop the soil from drying out. Alternatively, you can place the tray in a heated propagator, which will get the seeds off to a super start. Once the seeds have germinated and the seedlings are looking strong, remove the tray from the bag or propagator but the responsibility to water is now on you.

As the plant starts to grow and get established, plant the fibre pot into a bigger pot of compost and the roots will reach through the fibre-pot wall as the plant continues to grow.

RIGHT Fresh herbs in the kitchen.

IDEAL FOR INDOORS

These crops don't need pollination before harvest as we are just eating the leaves.

- **Soft herbs** like basil, parsley and coriander are annuals that must be sown every year.
- **Perennial herbs** like rosemary, thyme, sage and chives will see new growth every year once they're established.
- **Salad leaves** are great – why not sow a variety of seeds and pick off leaves as they get big enough to eat.
- **Spring onions** grow well indoors.

WINDOWSILL WONDERS

Here we have some compact plants from the Solanaceae family. These will all flower before fruiting and will need a visit from bees or other pollinators. If the plants are on an inside windowsill, remember to open the window on sunny days when the plant is in flower; alternatively, if there space to grow plants on the outside windowsill, maybe in a long trough, you grow the plants outside and open the window to water and harvest your food.

These suggestions are all annuals and need sowing in the spring for a summer harvest.

- **Tomatoes** – cherry tomatoes on a bush plant are a great option.
- **Chillies** – choose your favourite variety.
- **Aubergines** – look out for different colours and shapes.

CITRUS IN THE SITTING ROOM?

My son recently bought me a lemon tree for my birthday and, although it was a very small tree, it already had fruit on it. The instructions explained that this was a perfect house plant as long as we pop it outside for pollination when it flowers. I can't wait to slice my home-grown lemons into a gin and tonic.

POTS FOR PATIOS AND BUCKETS FOR BALCONIES

If you do have a little bit of outdoor space then consider bigger containers that don't need to be moved around for pollination. An easy crop with a prolific harvest is courgettes – try the round variety as the fruits are more compact. Carrots are now available in a rainbow of colours; try sowing spring onions in the same container so that as one crop grows down, the other grows up.

What about growing vertically? Tall sweetcorn, climbing beans, cucumbers and even melons. Provide a support, tying the stems in loosely, and your plants will reach for the sky.

Many dwarf trees are happy in pots and planters so search out your favourite fruit and you could soon be picking sun-kissed apricots or perfect pears.

It's important to remember that, even though these planting containers are a bit bigger, they will still need regular watering and topping up with fresh compost for nutrition.

HOME-GROWN SUPPER SALAD

From the windowsill
to the table.

INGREDIENTS

For the salad

Use whatever quantities
you like of . . .

tomatoes

sprouting chickpeas

spring onions

soft herbs such as
    coriander, chives
    or basil

For the dressing

1 tsp honey

1 tsp mustard

3 tbsp olive oil

1 tbsp white wine vinegar

salt and pepper to taste

METHOD

1. Cut the tomatoes into bite-size pieces and place in a bowl
   with the chickpeas.
2. Slice the spring onions, tear the herbs and scatter
   everything over the tomatoes.
3. Combine the dressing ingredients – I shake mine in a
   clean jam jar.
4. Drizzle the dressing over the salad and fold through with
   a large spoon before serving.

HOME GARDEN

'What shall we grow this year?' was the question that my friend Matt asked me over breakfast on New Year's Day after the previous evening's celebrations. I was pleased to see that he was eager to begin his plans to spend more time at his family's allotment and was impressed to hear of his successful iceberg lettuces and the bumper crop of pumpkins that his daughter, Matilda, had grown previously. Giving good time to planning your plot and choosing your crops can make all the difference when it comes to harvest time.

Each year, I record everything that we produce in our home garden at Hen Corner on our Harvest-ometer online calculator and therefore know that, between the fruit, vegetables, honey and eggs, we gather in food that saves us more than £250 per month on our food bill. With this under our belts, we are always ambitious to improve in years to come . . .

So how did I answer Matt's question of what to grow? Well, I suggested a few categories that we've incorporated into the growing spaces of our London back garden that should ensure his (and your!) toil in the soil is as rewarding as possible:

- **Favourite foods** – See what grows well in our climate (not bananas or pineapples in the UK!) and try growing something that you love to eat. French beans are fun, tomatoes are tasty and cucumbers are crunchy.

- **Best fresh** – Most of the food we buy in the UK is at least a few days old, if not older, and many fruits and vegetables start converting their natural sugars into starch once harvested. New potatoes and baby peas are just a couple of crops that taste amazing when freshly picked. Where else can you eat food within minutes of its harvest?

- **High value** – Raspberries and redcurrants can be pretty pricey in the shops but growing your own can produce bowl after bowl for many months of the year. Now that the squirrels have stopped stealing our figs we can harvest around £500 worth a year!

- **Perennials** – The old faithfuls that feed us year upon year: rhubarb, artichokes, hardy herbs, horseradish, etc. Our favourite is asparagus; once established, a bed can feed a family every week from April to June. We harvest around £30 worth of asparagus each year and it is so fresh and tasty.

LEFT Feeding hens in our London garden.

- **Fruits** – These usually harvest in season with the minimum of maintenance. We have apples, pears, peaches, figs, plums, apricots and a variety of soft fruit. We also have an olive tree that yields a good crop (the olives need to be cured in brine to remove bitterness before eating them) and an almond tree that we are constantly trying to protect from the squirrels.

- **That which stores well** – Winter squashes keep for ages in a cool place, rhubarb and runner beans are happy in the freezer, and other foods can be preserved or fermented into your favourite tipple.

Once we've got our fruit and perennial plants in the ground, they will usually stay in place long term. Then we can start planning for the annual crops that need sowing afresh every year.

LEFT Our garden in summer.
ABOVE Figs ripening on the tree.

CROP ROTATION

Annual plants can be grouped together in categories, often with other similar plants:

BED 1 – ROOTS	BED 2 – LEGUMES	BED 3 – BRASSICAS	BED 4 – ALLIUMS
Potatoes	Peas	Cabbage	Onions
Sweet potatoes	Beans	Broccoli	Garlic
Carrots		Cauliflower	Leeks
Parsnips		Brussels sprouts	

We try to use a four-bed rotation system to prevent vital minerals being depleted from the soil and to help protect from diseases. Each year, as the crops rotate, everything moves up a place on the list – so (for example) in year two we'd sow legumes in the bed where roots were in year one, and onions would be planted after harvesting the cabbages.

Other categories, such as Solanaceae – which include tomatoes, peppers and aubergines – can be planted in borders or anywhere there is a bit of space.

Cucumbers like to climb high; they don't need lots of ground space but do need regular watering to help the fruits to grow well. Try planting them near a trellis or over an arch. Once you are successful with cucumbers, why not try and grow melons in the same way?

COMPANION PLANTING

While we generally follow the planting plans above, we've found that some crops like to grow well as companions. For instance, carrots are prone to carrot fly, but the smell of onions hides the aroma of carrots and confuses the flies. So, we plant out our onions in late autumn or early spring, and once they are established and sprouting we sow carrot seed in between the rows. The onions are ready to harvest first and then, once we've lifted them, there is more space for the carrots to swell.

Carrot seeds are very small and carrots can grow big, so I mix the seeds with some fine sand for sowing to help distribute them more evenly.

Three Sisters is a traditional way of growing sweetcorn, beans and squash in a compact space. The corn grows upwards, reaching for the sky and wind pollination for the growing cobs; the beans climb up the sweetcorn, twisting and turning for good anchorage; and the squash grows around the base, sprawling its large leaves to cover the soil – this restricts weed growth and water evaporation, allowing space for the large fruits to swell.

NEVER EAT STRAWBERRIES AT CHRISTMAS!

Well, only in jam. The fresh strawberries we get in the winter will have been either grown in artificial conditions or flown in from other climates. Strawberries are truly at their best

RIGHT Broad beans are one of the first crops of summer.

freshly picked and still warm from the sun. Other berries, however – be it blackberries, raspberries, gooseberries and every type of currant (red, black or white) – will all freeze perfectly and can be used throughout the year for breakfast, desserts or smoothies.

Once established, soft fruit are very easy to grow. Different varieties will be ready to harvest at different times, allowing you to start picking ripe strawberries in late spring and still have raspberries ready in early winter. Soft fruit are quite straightforward to propagate, allowing you to produce additional plants. Throughout the summer, strawberries send out runners that put down roots in nearby soil; you can also position a filled pot close by and press the new crowns into the compost. You can reproduce currant and berry canes in late autumn by taking cuttings from soft stems at the end of the canes and poking the cuttings in a ring just inside the diameter of a pot of compost. I cover the cuttings with a plastic bag to stop them drying out and store them in a dark shed over winter. By the next spring, you should start to see new leaf buds, which indicate that new plants are growing. As they start to get bigger, carefully transfer each new plant into its own pot.

PROTECTING CROPS AS THEY GROW

All plants start out as small seedlings and while we may propagate and initially nurture them inside, at some stage they have to stand alone in the garden. This is great on one hand, as they will have space to send out deep roots and grow tall, but there are many challenges that they could face as they develop.

For us, the biggest challenge has been our free-ranging hens. One scratch of a claw can lift several seedlings in one go, so we have developed a hoop-and-net system that sits over each of our veg beds and have found not only that these keep the chickens away (along with cats, pigeons, foxes and so on) but also that they prevent butterflies laying eggs on our broccoli (so no hungry caterpillars) while still allowing bees in to pollinate our beans.

JOBS THROUGHOUT THE SEASONS

SPRING

HARVEST

Remaining winter brassica, such as purple sprouting broccoli, and our absolute favourite – asparagus. Once established, an asparagus bed can feed a family every week from April to June.

The next crop of the year will be broad beans.

TIP Plant different varieties of asparagus and cut spears, from below the soil surface, before the asparagus beetle finds them.

JOBS

Sow seeds indoors, in a heated propagator or in trays of compost covered with a plastic bag. Plant peas and beans straight in the ground in their final destinations.

SUMMER

HARVEST

Anything that's ready!

JOBS

Stop cutting asparagus and allow the spears to grow into tall ferns; this helps the plants to photosynthesize and feeds the underground crowns for next year.

Water all crops, especially in the morning of a hot day.

AUTUMN

HARVEST

Sweetcorn and squashes, potatoes, carrots, autumn raspberries and orchard fruits.

JOBS

Planting broad beans now will allow them to establish before the black fly aphids find them in spring.

Onions and garlic also benefit from being planted ahead of the game before winter.

The asparagus ferns that have been growing tall since we stopped harvesting the spears at the end of June will now need cutting down under soil level, and the bed mulched with compost ready for next year.

WINTER

HARVEST

Kale and winter greens will still be growing and parsnips will be ready to lift.

JOBS

Once most crops have been harvested, we give the plot a good clearing on a dry day. The grapevine always needs a good cut back, maybe in two stages – now and in spring.

Then we prepare the garden for winter: composting leaves, giving the raised beds a final weed and mulching compost on all empty beds.

Fruit trees benefit from a winter prune while they are dormant. You can remove all unwanted branches to ensure that there is space for light, air and swelling fruit next year.

ROASTED VEGETABLE TART

Easily adapted for peppers, squash or aubergines.

SERVES 8

INGREDIENTS

For the pastry
200g/8oz plain flour
100g/4oz margarine
a pinch of salt
50ml/2fl oz cold water

For the filling
2 courgettes
1 large onion
1 tbsp oil or margarine
150g/5oz cheese, grated
4 eggs
200ml/7fl oz cream

METHOD

1. Preheat the oven at 200°C/400°F.

2. Peel and slice the onion into thick wedges. Chop the courgette into 1cm/½ inch half-moons and put in a roasting tin with the sliced onions. Add the oil or margarine, season with salt and pepper and then roast for 15 minutes.

3. Make the pastry: rub together the margarine and flour in a mixing bowl, then add a pinch of salt and just enough cold water to bring it together in a ball.

4. Roll out the pastry on a lightly floured surface. Lower it into a dish or tin and prick the base of the pastry with a fork. Bake for 10 minutes.

5. In a jug, mix together the eggs and cream, then season with salt and pepper.

6. Scatter your baked pastry case with the roasted vegetables and grated cheese, then pour over the egg mixture.

7. Bake again for another 10–15 minutes, until the centre of the tart is light brown and firm.

PLANNING THE PLOT

Before we moved to the corner house, I put my name on the list for a local allotment plot. I knew exactly where I wanted it – walking distance from the house and edging the Grand Union Canal. I was offered a plot a bit further away, but politely declined as I knew I wouldn't be able to tend it properly and thought that a thirty-minute car drive was counterproductive when I was trying to increase our sustainability. I already knew a few of the plot holders locally and was keen to join their ranks.

Allotments are nothing new. What we see today originated back in the seventeenth century, when common land was divided up and allotted to individuals, allowing them to grow food for their families. Later, the Smallholding and Allotments Act (1908) made it compulsory for local authorities to provide allotments where there was demand for them. After a short decline in interest, the Dig for Victory campaign of the Second World War reinvigorated many to try their hand at growing their own food. But once rationing was over and cheap mass-produced food was easily available, allotments fell out of fashion.

Fortunately, we seem to have come full circle and today people seem keener than ever to get their hands dirty, quite literally. In London, due to shortage of space, waiting lists for allotments can be between ten and twenty years – probably because, once people have got their hands on a plot, they discover that the rewards reaped are well worth all the work. When you also consider the huge building developments in many urban areas, the opportunity to cultivate a little bit of land that you can call your own is becoming more and more appealing. As towns increase in size and new people move in, local groups and associations are key to building a sense of community; and if clubs and societies are established around common interests, such as growing your own food on an allotment, so much more the reward.

I was on the waiting list for fifteen years.

I received confirmation that I could have the end plot, Number 9, on Valentine's Day 2020. I couldn't believe it. I was so excited – the opportunities ahead were huge! Andy wasn't so thrilled. 'How will you find the time to look after it? There's lots of jobs that need to be done here in our garden. You can't do it all; something is going to have to go . . . '

'I'll keep it simple,' I reassured Andy. 'Fruit trees and honey bees.' I wasn't giving it up after waiting so long for it – it was going to be great.

I'm not sure which was the deciding factor – having an old friend on the plot next door or an opportunity to fish from the riverbank at the end of the plot – but Andy has certainly put the hours in, especially with the shed assembly and building raised beds. And now that

LEFT Our canal-side allotment.

we have a stove and kettle for tea, Andy is happy to come and potter at a weekend.

I felt really privileged, as though I was stepping into history, as I would be carrying on the traditions of food growing in our area that go back hundreds of years. From the sixteenth century, our town of Brentford and the surrounding area had significant market gardens, with 3,000 acres of land growing fruit and vegetables for the growing population of London. Strawberries grown in Brentford even won prizes at the Great Exhibition of 1851 in Hyde Park planned by Queen Victoria's husband, Prince Albert. So, with this as a heritage for my little plot of land, I was keen to lean into the legacy and had high hopes for a bountiful harvest.

While each plot holder has their own space, the sense of community is really encouraging. Walking past all the other plots to reach mine at the end, it's great to see what is growing and it's lovely to chat and catch up. I've found it inspiring, with others happy to share skills and experience and sometimes practically helping to get the job done. I've found their support and manpower invaluable – especially in the early days when I was starting from scratch.

OR SHOULD I SAY SCRATCHES?

The plot I inherited had been empty for around five years and the strong brambles had grown to 2.5m/8ft tall. I started with secateurs, snipping low down, then dragged the long canes to a pile at the back of the plot. While we could see the large fig tree in the centre of the plot, it was only once we started clearing that we found two pear trees choking in the middle of it all, straining through the forest of brambles to catch a glimmer of sunshine. We also discovered two apple trees and a greengage tree at the edge of the plot, all in need of a good prune. After removing the unwanted vegetation we were able to see the whole space and start to plan how we were going to use it.

I'd promised Andy that I would try and keep it low maintenance, and perennial crops that fruit every year can often bring a good harvest for minimum work. As the allotment was giving us much more space than our back garden, we could also set aside areas for crops that require more space, such as potatoes, onions and large squashes.

PLANNING OUT YOUR SPACE

With a newly cleared plot, or inheriting a more established space, it's good to decide on your aims and priorities and plan what to put where. This is what we did.

Assess what's already in place
Aside from brambles, rubble and buried plastic – I dug up a child's paddling pool! We found mature trees, some in the middle of the plot and others at the edge.

Water supply
Not only do we have push-tap standpipes between every three or four plots, but the plots edge the river so plants can also drink up through the soil. Think through how far you may need to carry a heavy watering can.

Shed

We chose a northern corner of the plot for our shed, so that it wouldn't cast a shadow over the growing space. We set it slightly back from the river so that it couldn't be seen too clearly from passing boats; and after clearing the ground, laying a good foundation of paving slabs and then assembling it, we painted it a pale green to blend in with the foliage behind. Inside, we installed shelf units on either side and use it for storing tools, seeds, beekeeping equipment, our camping stove and sheltering from the rain.

Fruit trees

I've planted two new trees to line up neatly with the uncovered pear trees. The first is a quince tree – I truly think quinces are the fruit of royalty, perfect for making jellies, desserts and the Spanish delicacy membrillo. The second is a Victoria plum – Andy's favourite and another reason to tempt him along with me. When planting new trees, there are many options available in terms of both variety of fruit and style of tree. For apples alone, flicking through a catalogue will reveal not just cooking or dessert apples but also bitter crab apples (brilliant for jelly), brown russets, Red Princes, the deep-green Bramley and everything in between.

One fruit farm in Kent, Brogdale, grows over two thousand varieties of apple – so when you're choosing a tree to plant, don't be limited to the fruit found in the supermarket; choose something really special. If you are spoiled for choice, you might consider a new 'Family Tree' where three, or more, cultivars are grafted on to the same tree, allowing a few branches of each type of apple.

If space is tight, apples can be grown as cordons along a fence, espalier fans against a wall, or low 'stepovers' as hedges. Dwarf bush trees can be even grown in a pot but do make sure they're regularly watered.

Permanent beds for perennials

Plan where you want your permanent crops, as they'll hopefully be there quite a while. All of ours are in the back half of the plot, allowing them to send down deep roots to drink well from the soil. Up behind the back path we have hardy herbs: thyme, rosemary and oregano, with wild garlic growing behind. In large rectangular beds framing the central fig tree we have our soft fruit: four beds of strawberries, one bed for blueberries, one for currants (black, white and red) and a row of rhubarb on the other side of the tree. Alongside the far path through the plot, we have rows of raspberries and gooseberries.

Beds for annual crop rotation

This leaves the front space for annual crops that need planting up each year. We've got twelve beds, allowing good space for single crops. In most years, we'll give a whole bed each to potatoes, runner beans, onions and carrots grown together, squashes and courgettes. I'll then use four on a rotation of alliums, legumes, roots and brassicas (see page 60). These front beds are nearest the river and, once our crops are established, they can often access all the water they need from the moist soil just a spade's-depth down.

Beds for a plot buddy

A good friend, also called Sara, was keen to help; so rather than set her to work for me, I've given her the first four beds on the waterfront so she can grow what she wants and take full responsibility for it.

A whole plot can be a lot to manage, especially if you are working at a day job as well. So, sharing the space makes perfect sense, along with sharing seeds and tips to grow high-quality, often heritage crops that just can't be bought in a store. An extra pair of hands to help with watering and harvesting when you are on holiday is invaluable, as is sharing produce and planning together.

PROTECTING CROPS

While growing food that is good to eat, we'll often discover that we are feeding more mouths than we intend to. Other plot holders can tell you what the main predators are and then all we can do is plan well – forewarned being forearmed – and plant more than we need as we know we won't get to keep everything.

From my experience, this is how we protect from the following:

- **Pigeons** love to peck at our brassicas. We cover all cabbages, broccoli, sprouts and so on with a wire cage, either makeshift from chicken wire or a more substantial netted frame that we can move from bed to bed.
- **Other birds** may fancy a feast on our fruit and I'm sure many berries originated as nature's way of feeding them. But we've found that fruit cages are the most reliable way of keeping fruit safe and hungry beaks away.
- **Foxes** seem to enjoy digging in recently turned soil, maybe out of curiosity or simply because loose soil is easier to snuffle in for tasty treats. For a week or so after planting, I try to cover beds with a wire mesh that the foxes can't scratch through.
- **Mice** like peas and beans just as they soften in the moist soil before germination. You can try starting them off inside, planting out when they've started growing and are a little stronger.
- **Slugs** love to hide in damp corners ready to slide out at night for a midnight feast. Keep the inside edges of your beds tidy, so they have nowhere to hide. Another trick is to tempt them with a beer trap – a plastic tub half-buried in the ground and topped up with ale. The slugs will visit for a drink and, if it's deep enough, will drown in it.
- **Butterflies** may look pretty, but don't forget that their babies are caterpillars –

very hungry caterpillars. Again, it's the brassicas that they'll be after, laying their eggs on the underside of the leaves. You can either go through every leaf regularly to wipe off the eggs, or, if you are already protecting the crop from pigeons, throw a net with smaller holes over your brassica cage to keep the butterflies out.

- **Codling moths** are the mothers of the maggots that you find in apples; they lay their eggs on the young apple leaves and fruit. We battle against these with traps, putting sticky gum around the tree trunks, and by spraying with organic nematodes in October.
- **Squirrels** seem to be the craftiest of them all. They are also very fickle and may change their minds, year on year, as to which crop they plan to rob. I've built a fort around my almond tree at home as I'm determined to eat them fresh in our London back garden, but I don't think permanent scaffolding around every tree is very practical or sustainable.

ABOVE Covering crops to protect from birds.

POLYTUNNELS FOR A LONGER SEASON

Some say that with a polytunnel their crops start a whole month earlier and can extend another month at the end of the growing season, so with the extra space that an allotment offers, it might be worth thinking about one.

While the plastic covers trap the heat in, they also keep the rain out, so regular watering is crucial; and in the summer they can become very hot and need doors open at both ends for ventilation. A big polytunnel is difficult to move, so if you are planting in the ground you must factor in a crop rotation by planting different categories in different sections of soil.

POLLINATION

Our secret weapons for the best fruit harvests are the beehives sited only metres from the trees. Even if it's raining, when the trees are in blossom our faithful pollinators make a dash for the nectar and trigger the magic of pollination as they gather in precious food for the hive. Neighbours who live streets away have commented on their increased crops since we've kept bees. It's great to hear how they are sharing the love. As the pollinated fruits start to swell, we watch patiently, secretly planning to pot up the pears in spiced syrup, get the greengages into jam and save up the apples for our cider-making.

SHOWING OFF

As the growing season comes to an end and we've picked, traded, eaten and preserved our harvests, we may decide to enter some of our finest produce into a local allotment show. In the same way that the rural farming community gathers for conferences in the winter and seed sales in the spring, late summer is definitely the time to show off your successes and lay out your harvest in front of the judges. A bit of competition often helps fuel enthusiasm too, especially if there's a prize in the offing. While I've previously enjoyed big county and agricultural shows, a local allotment show is perfect for those producing food in towns and cities. I entered a few classes in my first year as a plot holder and won a silver cup for my sweetcorn during lockdown!

In previous years, I've been honoured with the invitation to judge the 'domestic classes' at the annual show of local London allotments. The competition at our local show is divided into three main categories allocated to separate judges: fruit and vegetables, flowers and plants, and all the products you can make in the kitchen with the wonderful produce that you've grown. When I was judging this last category, I counted around thirty classes within it, each with up to five or six entries that I had to inspect, taste and place. There were several subcategories of pickles and preserves, home-brewed wines and beers, bread, cakes, tarts and pies, all rather delightful.

I remember a conversation with a plot holder at one of these shows who explained to me that she treats her allotment as a job, working hard to produce good food in the same way that she works hard in the office to pay the other bills.

Throw in the fresh air and exercise and the benefits just roll in.

SPICED BUTTERNUT SQUASH SOUP

Make a big batch and freeze in portions for an easy supper.

SERVES 4

INGREDIENTS

1 butternut squash

1 tbsp olive oil

salt, pepper and chilli flakes to taste

1 vegetable stock cube

500ml/1 pint boiling water

METHOD

1. Preheat the oven to 200°C/400°F.

2. Peel and chop the butternut squash, then roast in a tin with olive oil, chilli flakes and seasoning until soft and starting to brown, approximately 30 minutes.

3. Mix the stock cube with the boiling water in a large saucepan, then add your roasted squash and simmer until very soft.

4. Mash or puree the soup to your preferred consistency, adding more hot water if needed. Serve with a dollop of yoghurt, a drizzle of olive oil and a grind of black pepper.

FORAGE FOR FOOD

From newborn babies to teenagers home from school, an office worker on their lunch break or a grandmother searching her handbag for the last sweets in a packet, we all have a natural response to search for food when we are hungry. But is it true that there's no such thing as a free lunch?

Before humans settled in communities to grow food, they hunted and gathered. They chased animals to roast and picked fruits, nuts and vegetables throughout the seasons. Some treasures were high up in trees, some hidden within spiky bushes; edible tubers were discovered beneath the ground and some food was easy pickings from low-hanging branches. If you were hungry, you searched for food, as we still do today, and there's lots of great stuff to be gathered in.

NATURE'S LARDER

If you haven't been able to grow your own fruit and vegetables, you can still join the preserving party by taking a walk around your local area and foraging in nature's larder.

Every morning in late spring while brushing my teeth, I spy out of the window at the elderflower tree in my neighbour's garden. First in leaf, then clusters of buds and finally open blossoms of fragrant blooms that are perfect for a floral cordial. Once I've seen their tree display its creamy white flowers, I can be confident that there will be plenty more around the area yielding precious nectar to be enjoyed by the bees and foragers alike.

Later in the year, even in wet summers, the fruits of our hedgerows seem to be happy soaking up the rainwater and filling their berries with juice. As long as spring is warm and dry enough for the bees to fly and pollinate, blackberries, crab apples, plums, loganberries and the traditional mulberry should be bursting with reward for those on the lookout with a basket to hand.

Our towns and cities are full of rewarding foraging spots. Locally, my favourite place to find such treasures is along the Grand Union Canal from Brentford to Hanwell. Taking a couple of bags on a family cycle ride is all you need to add a bit of foraging to some fun exercise in the open air. Berries can be enjoyed straight off the brambles, with sticky blood-red juices incriminating your face and fingers; or, if you have the willpower to bring them home, they can top a cake, be frozen for later, or preserved in a jelly or jam to be enjoyed on scones, toast or stirred through a winter rice pudding.

LEFT Crab apples make lovely jelly.

The general rule for foraging is to ask permission of the landowner. If you find something you fancy on a public footpath you should be absolutely fine to gather fruit under common rights law, providing it is growing wild and is for personal use only. Just watch out for Sites of Special Scientific Interest (SSSI), many of which are on private land with no public access, as they forbid the 'removal or damage to any plant, fungus or animal'.

While some people are concerned that foraging is not good for the environment and natural habitats, let's remember that picking fruit doesn't damage the plant itself. Fruit is the way that plants reproduce, and in wild hedgerows there are usually plants in abundance. Just don't dig them up to replant them at home! Other people are passionate about leaving fruits and berries for wild birds, and I agree; however, in my experience, the local wildlife seem to prefer to forage just before the fruit is ripe enough for us and will often snack from the higher branches that are out of our reach.

Locally, we have a fabulous organization, Abundance London, that maps fruit trees that are often left unpicked or have become too much of a challenge for their owners. Volunteers are coordinated to pick this fruit for their own use, allowing those without growing space to enjoy their share of the urban harvest. Maybe look out for similar groups in your area?

ABOVE Fresh ripe blackberries.

WILD GARLIC PESTO

Pesto is a fresh-tasting sauce for pasta or a delicious dip for freshly baked bread.

While we tend to think of it as made with basil, any number of soft leaves can be ground together with oil, adding nuts and cheese as we choose. Nasturtium leaves, spinach and wild rocket make good alternatives; remember to add in a few cloves of garlic.

MAKES 1 JAR

INGREDIENTS

1 large handful of wild garlic leaves

70g/2½oz Parmesan cheese, grated

juice of 1 lemon

200ml/7fl oz rapeseed oil

a pinch of salt

METHOD

1. Put everything except the oil in a food processor and blitz to an even paste. Alternatively, you can use a pestle and mortar.

2. With the motor still running, slowly pour the oil into the paste, watching as the rich sauce takes shape.

3. If you don't eat it all in one go, pop into cold sterilized jars (see instructions for sterilizing on page 92), cover the pesto with oil, seal and store in the fridge. You can also freeze it in little bags.

ELDERFLOWER CORDIAL

Summer in a glass. With such a short time in bloom, we try to make enough each May to last the whole year.

MAKES 10 LARGE BOTTLES

INGREDIENTS

40 elderflower heads

4 lemons

170g/6oz citric acid

5kg/10lb sugar

3 litres/5¼ pints water

METHOD

1. Give the flowers a good check over, shaking off any small bugs that may be hiding in the blooms. Place in a large container – we use a fermentation bucket for this.

2. Pare the zest off the lemons with a potato peeler before cutting the lemons into 1cm/½ inch slices, then add both zest and slices to the flowers in the bucket.

3. Sprinkle the citric acid over the flower heads and lemons.

4. Make a strong syrup by gently warming the sugar and water in a large pan until fully dissolved. Bring the syrup to the boil for 1 minute.

5. Pour the hot syrup over the ingredients in the bucket, then cover and leave to infuse overnight.

6. The following day, strain the mixture through a fine cloth, muslin or tea towel into a large container. Compost the used flowers and lemon.

7. Decant the finished cordial into sterilized bottles (see page 92), then seal and label.

JELLIES

Apples, be they freshly picked, windfall or even crab apples, make a great base for jams and jellies as they provide the pectin needed for a good set. If you find crab apples, which are much too sour to eat raw, they just need to be quartered and simmered in lots of water before you strain the precious juice to boil up with equal quantities of sugar, creating a jelly base for many different flavours. Plain crab apple jelly is nice, as the sugar offsets the sour fruit flavour. Rosemary or thyme sprigs can be added to create an accompaniment for roast meat, or you can include other fruit found on your forage, such as blackberries or greengages, for a simple jam.

CORDIALS

Fruits, flowers, rosehips and herbs all make great base flavours for cordials, which can be stored throughout the year ready to serve diluted with chilled tap water, sparkling water or, for a special treat, sparkling wine.

STORING AND PASTEURIZING

Cordials can be used straightaway or stored in the fridge or freezer to prevent the natural yeasts from fermenting. If you are making larger batches, you can pasteurize the cordials to kill the yeast once you've decanted them into strong glass bottles, which will allow their safe storage at room temperature.

I pasteurize my bottles of cordial by placing a folded tea towel in a large saucepan and carefully positioning the sealed bottles on top. I then top up the saucepan with water and bring it to a gentle simmer, holding it at 75°C/167°F, with just a low heat, for twenty minutes. I then turn the heat off and leave them undisturbed to cool down naturally, then store in a cool, dark place.

FRUIT LIQUEURS

Making fruit liqueurs can be a fabulous annual event: grab some friends, forage for fruit and start planning your Christmas presents. This is a very flexible recipe that guarantees a special treat as an aperitif, cocktail ingredient or after-dinner drink.

MAKES APPROXIMATELY 70CL/1¼ PINTS

INGREDIENTS

250g/8oz autumnal fruit such as damsons, plums, sloes, raspberries, blackberries

125g/4oz granulated sugar

375ml/13fl oz spirit such as gin, vodka, whisky, rum, brandy

METHOD

1. Pick over the fruit, removing stems, leaves or damaged fruit.
2. Place in a suitable bag or container and freeze overnight to soften the skins – this helps release the fruit juices.
3. Tip the frozen fruit into a large glass or plastic container with a tight-fitting lid. Sprinkle the sugar over the fruit and top up with the spirit of your choice.
4. Seal well and leave at room temperature for a couple of months, giving the jar a gentle sideways turn each day to ensure even infusion.
5. Decant the finished liqueur into pretty bottles and save the drunken fruit for a special dessert or grown-up cake.

PRESERVE

 Patience is a virtue, and all good
things come to those who wait –
both very true when it comes to
pickling and preserving . . .

Once we've started growing our own food, the next stage is
preserving the harvest so that it lasts as long as possible, allowing
us to continue eating it throughout the year.

Here at Hen Corner, once we've gathered in the harvest, plums
and berries will be made into jam and vegetables from the kitchen
garden will be transformed into pots of pickles and chutneys.

I love preserving – picking something wonderful in its moment of
perfection and doing all you can to maintain its goodness for as long
as possible. While there are many techniques to choose from, the
mission is the same: to prevent the deterioration usually caused by
bacteria. Germs need warmth and moisture to breed, so by reducing
the water content of our food through salting, cooking and drying,
we decrease the chance of it spoiling. Freezing food inhibits
bacterial growth (it's just too cold) and high-sugar/acid/alcohol
environments similarly prevent bacteria from thriving.

Sometimes I think I prefer the preserved goodies to the fruit and
vegetables when they are freshly picked. Patience is a virtue, and all
good things come to those who wait – both very true when it comes to
pickling and preserving. We might harvest in the sunshine, but most
preserves need time to mature so they are ready to be enjoyed a few
months later at a midwinter feast. Label well and squirrel them away
in a dark place to enjoy when the tree branches are bare and the
nights are drawing in. If you haven't been able to grow your own, look
out for local produce in shops and farmers' markets – buying it in
season may well be cheaper and is much more sustainable.

DRYING

When we hear the term 'dried food' it can be easy to think of old, stale items that are well past their best. But if we know that bacteria needs moisture to grow, then we can understand that drying food is the proven way to protect our harvests and allow us to store them safely for the rest of the year.

Sometimes I've found food drying more by accident than design, but it's still valuable and can be used in future recipes. For example, if I've got fresh herbs and mushrooms, I store them in the fridge in either a cardboard tray or an open paper bag; this allows moisture to naturally evaporate and reduces the opportunity of bacterial growth. We can then use them fresh if we want to and any that aren't used will slowly dry out. Once completely dried, the items can be stored in clean airtight containers, such as old jam jars, and added into recipes when needed.

If you've dried mushrooms this way, they will reduce in size and feel a lot lighter – but soak them in warm water and they'll soon soften and can be chopped up and added to risottos and pasta dishes.

Many herbs are easy to grow, with or without a garden, but not all of them can be harvested all year round. When herbs are at their peak, cut and tie them in bunches with garden string. Hang them upside down in a warm kitchen and soon you'll have dried leaves that you can crumble by hand, store in an airtight jar and use as desired.

Tomatoes are relatively easy to grow, but when they are ready we sometimes find that there are more than we can eat. Many countries have the climate to dry their tomatoes in the sun; if our weather is not as reliable, try drying your home-grown tomatoes on a tray lined with parchment in a low oven.

Dried fruit has always been popular and if you've got lots of fruit trees, you might want to invest in or hire a dehydrator. While we may immediately think of raisins, apricots and cranberries, these are not commonly grown in the city. Bags of dried apple rings or home-grown figs will not only make a delicious snack but could also be a beautiful gift.

LEFT Dried apple rings.

FREEZING

As a child of the 1970s, I remember the excitement of our first freezer – a huge chest that could store tubs and tubs of ice cream, large joints of meat and spare loaves of bread. I remember my mum doubling up on her baking each time she made a cake so that one could be sealed in a plastic bag (with the air sucked out through a long drinking straw) and safely preserved for later in the freezer.

As freezers became more popular, so did processed food that could be cooked quickly, tipped from colourful packages on to oven trays for a fish-and-chip supper in minutes. While frozen pizzas, chicken nuggets and ready meals are extremely convenient, I hope you know me well enough by now to be confident that I'm not recommending TV dinners that you reheat in the microwave. However, while still a relatively new process, freezing is a most convenient way of preserving food, especially food that we've grown ourselves.

We are fortunate to have space for a chest freezer in our garden shed and find it really useful for storing spare bread from our weekly micro bakery, bags of home-grown fruit and veg, and various cuts of meat that we buy, by the whole animal, direct from the farm.

If we've spent months nurturing home-grown crops, the quickest way to preserve many of them, come harvest time, is in the freezer. By following a few top tips, we can ensure that they stay in perfect condition ready to grab for another meal throughout the year.

PREPARING VEGETABLES FOR FREEZING

- Choose vegetables in good condition; trim off anything past its best.
- Peel and trim ready for cooking and cut into bite-size pieces.
- Cut broccoli and cauliflower into florets.
- Pop large beans, such as broad beans, and peas out of their pods.
- To retain good colour and texture, blanch the prepared vegetables in boiling water for just two minutes, then cool quickly in cold water to stop further cooking. Drain well, transfer to a clean plastic bag or container, then seal and freeze.

PREPARING FRUIT FOR FREEZING

- Peel, remove core, then chop or slice apples and pears, before sealing in a bag to freeze.
- Open freeze delicate berries on a tray until they're firm, then tip them into a rigid container, seal and freeze.
- Plums – to stone or not to stone? It doesn't really matter. Obviously, it's quicker to

simply rinse and drain plums before freezing whole; just remember that they still have stones if you are planning to bake them in a pie. Alternatively, cut them in half and remove the stones before freezing. Freeze in portion sizes or open freeze, like berries, until firm.

- You can also cook fruit down with a little sugar and freeze in tubs ready for a quick pie filling or compote to serve with yoghurt.

ABOVE Frozen blackberries, raspberries and blueberries.

JAM

For many years, most developed countries have relied on importing a significant percentage of the food we eat, which has led to the expectation that our favourite foods will be available all year round. But why are we surprised when we eat strawberries in the winter and they don't taste 'like they used to'? Just take a minute and think about where those strawberries have been growing during the coldest months, when they were picked, and how they have travelled to your plate – a completely different experience to the dark-red ripe fruits that you pick yourself in June.

As we are trying to produce as much of our own food as possible, our family life revolves around the seasons, and I see the tasks and responsibilities of each year broken down into clear lists that need to be addressed in line with nature. Some of our jobs are weather dependent, some rely on ideal temperatures, and all weave into the tapestry of tasks that help create the best environment for a garden that will yield a healthy harvest. When strawberries are in season, it's time to make jam!

While we might not fancy a steamy afternoon in the kitchen, boiling up pans of fruit and sugar, when the sun is shining hot at midsummer, the reward of a homemade cream tea is often the only incentive that I need to get my big saucepans out. Strawberries are hard to resist when freshly picked and sunshine-warm; they are wonderful in desserts, on cakes or just in a bowl with cream. But if you've got a glut, have visited a pick-your-own farm or have even found them on special offer in the supermarket, strawberry jam, with whole fruits, is really the nicest homemade jam you can make.

Even though I make hundreds of jars of preserves each year, I have nowhere near the output of Thomas William Beach, the 'father of the jam trade', who started growing prizewinning fruit in Brentford back in Victorian times. His harvests were so remarkable that by 1867 he had taken a lease on an additional 26 acres very near our house, including the site that became Griffin Park (Brentford Football Club), and started preserving the bounty that wouldn't make the journey to market. Gathering up spare fruit grown by family and friends, they also made wine with blemished fruit not good enough for jam. He was a great local employer and his new factory site included both a billiard hall and theatre especially installed to lure his workforce away from the many pubs in the area (admission for a show was one old penny or two empty jam jars – this was a sustainable business ahead of its time). As he continued to win many awards for his excellent products, he didn't forget the local needy and supplied them with soup and bread during bad winters. What a hero. I'd have liked to have met him; I'm sure we would have had loads to talk about.

LEFT Homemade jam on warm scones with cream.

While my favourite way of enjoying jam today is with clotted cream on a scone or sandwiched in a Victoria sponge cake, jam was an important element in the diet of those living in nineteenth-century London. We think of bread and butter as a staple snack, but in those days dairy produce was much too expensive so treacle – a by-product from sugar refining – became the spread of choice on bread for extra energy and a sweet treat. When factory-produced jams made with real fruit, such as Beach's, became affordable, they were seen as a nutritious health product – which I suppose it was compared to treacle.

BASICS OF JAM

There are three key processes going on when we make jam.

- First, as we boil the mixture, we are killing the yeasts naturally present on the fruit to prevent fermentation.

- Second, again with the boiling, water is evaporated from the fruit, making it harder for bacteria to grow and allowing the fruit to stay good to eat for years.

- Third, the sugar absorbs the remaining water, creating an environment that inhibits bacterial growth.

I usually use equal quantities of fruit and sugar and add lemon juice to both balance the sweetness and add pectin to help the jam set firm.

SETTING POINT

For a good set jam that will hold its own on a scone, we need pectin to turn the juices into a jelly. Some fruits, such as apples and lemons, are naturally high in pectin, but if these are not in your recipe you can buy commercial pectin to add into the pan.

These are the reminders that I use to make sure that my jam is cooked enough before potting it up:

LEFT Just three ingredients: strawberries, sugar and lemons.

LEFT Potting up jam into warm jars.

- **Wait** – until you have a high foaming boil
- **Watch** – until the bubbles thicken and darken
- **Wobble** – a drop of jam on a spoon
- **Wrinkle** – a drop of jam on a cold saucer.

OTHER FRUITS

Jam is a great way of preserving most fruits, especially if you have grown them yourself. Following the same technique and quantities on the opposite page, I add mixed spice to my plum jam and the zest and juice of a couple of oranges to my fig jam.

STERILIZING JARS

It's really important that your jars are sterilized and sealed properly to avoid unwanted bacteria finding its way into your jam and spoiling it.

I wash my glass jars and their lids in hot soapy water and then rinse them well. The jars are then placed right way up on a baking tray and warmed in the oven at 120°C/250°F for at least twenty minutes. I keep them warm in the oven until I need them, as it's important to put hot liquids into hot jars to prevent the glass from cracking.

Meanwhile, I place the lids in a saucepan of simmering water so that they are ready to seal the warm jars as soon as they are filled. Preserves in sterile jars will keep for several years.

STRAWBERRY JAM

This can easily be adapted for other soft fruits or berries.

MAKES APPROXIMATELY 5 X 340G/12OZ JARS

INGREDIENTS

1kg/2lb strawberries, rinsed

1kg/2lb granulated sugar

2 lemons

METHOD

1. Remove the stalks from the strawberries and cut the fruit into half or quarters if they are very large, then place the berries in a large preserving pan. Sprinkle on the sugar and the juice from the lemons, reserving the lemon skins, pips and flesh.

2. In a smaller pan, boil the reserved lemon pieces in enough water to cover them for 10 minutes.

3. Meanwhile, gently bring the strawberries, juice and sugar to a simmer.

4. Strain the hot lemon water through a sieve into the strawberries and press the pulp through with a wooden spoon – this is high in pectin and will give your jam a good set.

5. Bring the jam to a high foaming boil and bubble rapidly for 15 minutes, until the setting point (105°C/221°F) is reached.

6. Leave the jam to stand in the pan for 20 minutes to cool a little and allow the berries to settle. Pot in sterilized jars (see instructions for sterilizing on page 92), then seal and label.

PICKLING

I f it's a firm vegetable that you want to preserve, pickling it in a sweetened spiced vinegar is a great way to go. This will keep the crunch of carrots, cauliflower and green beans rather than allowing the vegetables to stew down into the thick gloss of a chutney.

Pickles are perfect in sandwiches, as an accompaniment to cheese boards or simply as a spoonful on the side of your dinner. I was first introduced to piccalilli on childhood holidays in Belgium, where it was a popular alternative to mayonnaise to top a cone of frites.

Good pickled onions are wonderful. In fact a humble jar of pickled onions won Best in Show at a recent allotment show that I was invited to help judge. But, as always, if we are preserving what we have a bounty of, adapt the recipe to suit your supplies.

We always begin pickling with preparing and salting the veg to draw moisture out, and then prepare a sweetened vinegar to protect from bacteria and allow the spiced flavours to permeate through.

LEFT Pickling veg in the summer to store for winter.
RIGHT Mixed pickles to snack on.

PICKLED VEGETABLES

I've always loved crunchy pickled onions and am often disappointed with the flavour and texture of those bought at the supermarket. If you have grown your own or found a source of cheap vegetables, try this simple pickle.

MAKES APPROXIMATELY

4 X 340G/12OZ JARS

INGREDIENTS

1kg/2lb small onions, cucumbers, or other firm veg of your choice

1 tbsp salt (for overnight salting)

350ml/12fl oz vinegar of your choice

200g/7oz sugar

2 tsp mustard seeds

1 tsp coriander seeds

1 tsp cumin seeds

½ tsp peppercorns

METHOD

1. Prepare your vegetables by peeling and chopping as desired – I keep small onions whole (tops and tails removed) and cut cucumbers into long wedges the height of the jar.

2. Sprinkle the vegetables with salt, cover and leave to rest overnight as the moisture is drawn out.

3. Prepare sterilized jars (see page 92).

4. Dry fry the spices in a large pan over a low heat for a few minutes to release their oils and flavours, then add in the vinegar and sugar and warm to a simmer.

5. Rinse and drain the vegetables, then lower into the hot vinegar mix.

6. Gently pack the vegetables tightly into the warm jars, topping up with hot pickling liquid, seal while still hot and allow to cool slowly.

Many years ago, I was on a television show with the well-known food writer Xanthe Clay. When she overheard me explaining that we are self-sufficient in fruit and vegetables for several months of the year, she asked to come and visit to write a piece for *The Telegraph*. 'I'll bring a photographer – could you make a recipe for us?' No problem, I thought, looking around as to what was in season. We certainly had a good stash of veg, some fennel seeds were growing, and if I could add in some honey that would give a shout out to our bees.

Before she arrived, I gathered ingredients together on the garden table and decided that a piccalilli, preserving vegetables in a spiced vinegar-based sauce, would be of interest. As we chatted about the chickens, I began to put together the recipe. Actually, no, I didn't have a recipe – I was freewheeling it! So when Xanthe checked that I would be able to send her the specific quantities and method, my jaw dropped in horror – my goodness, I was making it up as I went along. 'No worries,' she said, 'I'll just write down what you are doing.' The following week, it was published across the country and it's been the recipe that I've stuck to ever since. Good job, too, as ten years later the Guild of Fine Food has just given our Garden Piccalilli a coveted Great Taste Award!

GARDEN PICCALILLI

Use whatever vegetables you have in plenty and that taste good raw.
A mix of runner beans, cauliflower, cucumber, carrots, shallots and
French beans works well . . .

MAKES 10 X 340G/12OZ JARS

INGREDIENTS

1kg/2lb fresh vegetables,
    chopped into small
    pieces

1 tbsp sea salt

1½ tbsp coriander seeds

1½ tbsp cumin seeds

1½ tbsp fennel seeds

1 tbsp mustard seeds

1½ tbsp ground turmeric

2½ tbsp cornflour

1½ tbsp mustard powder

500ml/17fl oz white
    vinegar

100g/4oz white sugar

4 tbsp honey

METHOD

1. Mix the vegetables with the salt and leave in a bowl
   overnight.

2. Drain the vegetables, rinse well and then drain again.

3. Crush the coriander, cumin and fennel seeds in a pestle
   and mortar. Mix with the mustard seeds, turmeric,
   cornflour, mustard powder and enough white vinegar to
   make a double-cream consistency.

4. Put the rest of the vinegar in a huge pan with the sugar
   and honey. Heat gently until dissolved, then bring to the
   boil. Add the spice paste and simmer for four minutes,
   stirring occasionally.

5. Take the pan off the heat and stir in the vegetables.
   Spoon into hot sterilized jars (see instructions for
   sterilizing on page 92), then seal and label.

While most recipes recommend keeping pickles for a
couple of months before using, if you can't resist eating
them straight away, we won't tell anyone.

CHUTNEY

While many fruits can be perfectly preserved in a jam, they can also pack a punch when combined with vegetables in chutney. A chutney is usually made with equal quantities of sugar and vinegar enveloping the most imaginative combination of fruits and vegetables (around three or four times the weight of sugar), flavoured with garlic and spices as you wish. Stone and orchard fruits are great pan fellows with soft vegetables like squashes and tomatoes.

As we know, fruits and vegetables are wonderful when fresh, but it doesn't take long for bacteria to grow, especially if there is moisture present as well as air. Making chutney disarms bacteria growth in several ways and, in my opinion, creates a fabulous new food product that is so much more than the sum of its parts.

HOW DOES CHUTNEY PRESERVE FRUITS AND VEGETABLES?

- Chopping the ingredients increases the surface area and adding sugar draws moisture out.
- Cooking kills any bacteria present in the ingredients.
- Cooking causes more water to evaporate; bacteria need that moisture.
- Vinegar creates an acid environment that is hard for bacteria to thrive in.
- Potting in sealed sterilized jars reduces air contact, and bacteria need air to grow.

So now that we have the basics, we can get creative with all kinds of recipes. First start with the fruit and vegetables that you have to hand. If you've been growing your own, I'm sure that there will be certain crops that have been more productive than others. For me, it's always courgettes. While I love them griddled and served with feta cheese, roasted for a couscous salad, sauteed in a tagine or even baked in a chocolate cake – we really can't eat all that we grow.

Another harvest that works well in chutney is the green tomatoes that just don't seem to ripen on the vine, or that you removed quickly because you feared that blight disease might take them. Simply add some chopped onions, garlic and some spice and cook down in equal quantities of sugar and vinegar. If you want your chutney dark with a caramel taste, use dark brown sugar and a dark vinegar such as cider or red wine vinegar. If you want the green colours to shine through, use white sugar and a light-coloured vinegar – it's up to you!

LEFT Sweet and spicy pumpkin chutney.

I'm a firm believer in utilizing what you have to hand and not buying in extra ingredients – so be flexible with what you have and get creative.

A few years ago, some visiting friends asked if I'd like some pumpkins. Their daughter, Matilda, had nurtured a patch on the family allotment to give out to her classmates but had discovered that many more had grown than she could give away at school. I hate food waste and I've never been able to turn away an offer of home-grown veg, so I created a recipe that combined her contribution with our glut of apples, a handful of raisins and some warming spices.

PREPARING THE FRUIT AND VEGETABLES

When preparing your fruit and vegetables, try to imagine what a spoonful of the finished chutney will look like. I find it's best to cut up ingredients into small pieces so that each spoonful has a little of everything in it.

- Apples and pears: remove the core and chop the fruit small – no need to peel.
- Pumpkins and large winter squash: peel off the hard skin, scoop out the seeds (these can be dried and roasted), then chop the flesh small.
- Stone fruit such as plums, damsons, cherries and so on: REMOVE THE STONES! You really don't want to break a tooth when eating a cheese and chutney sandwich.
- Onions and garlic: peel and finely dice.
- Tomatoes: cut them up small.

ADAPTATIONS

This basic recipe can be used to make chutney out of any produce (think soft fruit and vegetables) that you have in abundance. It works really well with green tomatoes that are unlikely to ripen fully, or even plums. Equal quantities of sugar and vinegar, along with your favourite spices, form the basis of a great preserve. If you want to include some dried fruit, perhaps raisins, dates or dried apricots, you may want to reduce the sugar by 100g/4oz or so.

COURGETTE AND CUMIN CHUTNEY

This is a classic chutney that is best left to mature for a few months, allowing the flavours to mellow as they are absorbed throughout all the ingredients. It makes a perfect treat for Christmas – to give away as a gift or to serve with cold meats and cheese on Boxing Day.

MAKES 8 X 340G/12OZ JARS

INGREDIENTS

2.5kg/5½lb courgettes

1kg/2lbs onions

100g/4oz cumin

6 cloves of garlic

4 fresh chillies or 2 tbsp dried chilli flakes

1kg/2lb soft brown sugar

1 litre/1¾ pints cider vinegar

METHOD

1. Chop the courgettes and onions into small, evenly sized pieces.
2. Finely chop the garlic and chillies.
3. Place all the ingredients in a large preserving pan and start off over a low heat.
4. Slowly bring to a simmer, giving it an occasional stir to ensure even distribution of sugar and spices.
5. Allow to simmer for 3–4 hours until the chutney has reduced in volume by half and the consistency is thick. Keep an eye on it throughout so that it doesn't burn or catch on the bottom. You'll know it's ready when you can draw a spoon through the chutney and see the bottom of the pan.
6. Pot up in warm sterilized jars (see instructions for sterilizing on page 92), then seal and label.

SALTING

S alting, or curing, is one of the oldest preserving methods, going right back to ancient civilizations. Discovering that rubbing food with salt, or plunging it into a salty brine, would protect it from spoilage was a real game changer for those original hunter-gathers. It must have been a huge relief to no longer have to eat the whole gazelle in one go as you didn't know where your next meal was coming from.

Once a necessity, now a delicacy, cured meats, fish and vegetables are often the centrepiece of an international smorgasbord. Think salmon gravlax of Scandinavia, Italian Parma ham, Greek olives and Moroccan preserved lemons. Cultures around the world have discovered that traditional techniques can produce wonderful creations with the bonus of a longer shelf life.

While I would hesitate at building a smokehouse at home, wrapping a side of salmon in a salt mix sweetened with sugar and flavoured with herbs and spices is really quite straightforward and well worth waiting the four days as moisture is drawn out, flavours permeate through and the fish firms up for easy slicing.

Many years ago, I'd bought half a pig from a farmer friend and I decided to cure our own ham for Christmas. Rubbing a huge organic leg of pork with salt and then storing it in our garden shed for a few months (in an airtight container for protection) felt really counterintuitive but it worked perfectly.

More recently, I was working in our garden room and spotted a cheeky squirrel helping itself to olives from our tree. Realizing that the fruit must now have reached peak condition, as the squirrel seemed determined to strip the tree bare, I decided to get in on the action and picked a whole kilo of firm green olives. After a quick bit of research, I discovered that a salt solution would not only help preserve this urban harvest but by drawing out moisture would also remove the unpleasant bitterness.

A beautiful recipe that is delightful to look at and delicious to eat is preserved lemons. This is a great way to utilize spare lemons if you've bought too many; and once the process is complete, a little goes a long way in flavouring Moroccan-style dishes. Preserved lemon chopped and stirred through rice or added to a tagine is often the secret ingredient that pulls the whole meal together.

PRESERVED LEMONS

Create your own Middle Eastern ingredient. This technique has been used for hundreds of years to bring a taste of summer to savoury dishes throughout the seasons.

MAKES 1 LARGE JAR

INGREDIENTS

9 lemons
4 bay leaves
200g/7oz salt

METHOD

1. Wash the lemons, trim off the tops and bottoms, and slice lengthwise into eight wedges.

2. Pack the lemons tightly in a sterilized jar (see instructions for sterilizing on page 92) with the bay leaves, adding the salt in stages as the jar fills. Seal and label.

3. Turn the jar to agitate it every couple of days. The salt will draw moisture out of the lemons and the liquid may bubble up initially.

4. After a week, if you see gaps of air around the lemon pieces, open the jar and use a clean fork to press the lemon down tightly to help it all become submerged in the brine. Add a little water, if needed, to ensure the lemons are covered.

5. The lemons should be ready to eat after a month.

FERMENTING

After many years making chutney and pickles, I was given a sample of piccalilli at a food festival and was curious when the stallholder told me that it contained no sugar or vinegar. 'It's fermented,' they said, which rather confused me as I'd always associated fermentation with making alcohol or, more recently, bread. How do you ferment vegetables? What is the pickling liquid that preserves them?

The vegetables that I was used to potting up were often soft courgettes, tomatoes and squash that would soak up a sweet chutney sauce, or crunchy onions pickled to perfection. I'd never considered preserving carrots or cabbages in a jar; I'd keep them in the fridge or pop them in the freezer. Then I remembered sauerkraut – that was preserved cabbage, wasn't it? While I'd never found the thought particularly appealing, I confess that I hadn't tried it often, maybe just once with sausages and beer on a trip to Germany.

This piccalilli, though, was something else. It was crunchy and zingy, fresh-tasting and certainly woke up my tastebuds. So I started looking into fermentation with an increased enthusiasm.

The simplest way to explain fermenting vegetables is that, instead of adding vinegar – which is often made from fermented fruits in the first place – we allow the natural juices in the vegetables to ferment into a liquor that becomes the preserving liquid. I've made cider enough times to recognize the carbon dioxide bubbling up within minutes of the pressed juice filling the demijohns. Natural yeasts are everywhere and love to feed on carbohydrate-rich liquids.

Chopping, slicing and grating increases the surface area of vegetables and salt draws liquid out by osmosis.

There are many books that you can read to get a comprehensive understanding of fermenting food (see Inform on page 170), but the following is an easy guide to getting started.

FERMENTED VEGETABLES

Fermented vegetables have many associated health benefits due to the good bacteria that can be cultivated, along with enzymes that can boost our immune systems and improve our gut health. So gather up some seasonal veggies and let your imagination run wild.

INGREDIENTS

Any combination of firm vegetables will do. We like apple, beetroot and ginger – or our 'plot rainbow' of radishes, courgettes, carrots and onions with chilli, garlic and ginger to taste. You can also make fermented piccalilli using cabbage, onion, carrot, red pepper, cauliflower and red chilli along with a few spice seeds such as cumin, coriander and mustard if you fancy.

METHOD

1. Finely slice your vegetables, then transfer to a large bowl and massage with a generous sprinkling of salt.

2. Loosely cover and leave overnight at room temperature. The salt will draw out the sugar-rich moisture from the vegetables, which will start to ferment.

3. By the next morning, the vegetables will be sitting in a puddle of juice. Pack them tightly into a large sterilized jar (see instructions for sterilizing on page 92) with enough 'seeped-out' liquor to cover them, then close the jar securely with an airtight seal.

5. Release the pressure a little each day for the first week as it ferments by loosening the lid to allow a little air out before sealing it tight again. Taste the veg at weekly intervals until the desired pickle taste is reached.

6. When you are happy with the flavour, store the jar in the fridge to slow down further fermentation. Once you've started eating this, push the remaining vegetables down to ensure they are covered with the fermenting liquor.

KEEP

> Keeping chickens in London? That's crazy! There's just not enough space. It would be fine if we lived in the country but you can't keep animals in your back garden – what will the neighbours think?

This is the conversation I had with Andy as we were trying to increase the amount of food that we could produce from our back garden. Yes, we had some veg beds, we were growing potatoes in a dustbin, tomatoes and onions in the flower borders and huge courgette plants with their brilliant yellow trumpets, but I really wanted to make one meal where everything came from the garden and I was pretty sure that my boys wouldn't see beans as their main source of protein, the centrepiece of the meal.

'I really think we need some animals,' I said, thinking back to the 1970s sitcom *The Good Life* and recalling Felicity Kendal wrangling some muddy pigs while looking stunning in denim dungarees. We could even rear a turkey for Christmas – wouldn't that be amazing?

With fruit and vegetables to hand, rearing animals allows us to produce a wider variety of foods. While there's more responsibility, there's also more reward. As people are becoming more hesitant to eat any products that come from an animal, the only way to be 100 per cent confident in how those animals are reared is to get involved ourselves.

CHICKENS

P eople often ask whether I grew up in the country and used to keep chickens as a child, but nope: I was born in Barking, brought up in East London and have been a city girl all my life.

It was only recently that my father told me of his childhood growing up in East London during the war and maybe this points towards my love of hens. He remembers his own father coming home one day with three baby chicks for him to look after. He was delighted with his new pets, oblivious to his mother's persistence that the animals were to be well fed and cared for with the sole aim of supplying an extravagant Christmas feast to rival the neighbours in the days of food rationing.

As the chicks grew, my dad and his younger sister naturally became fonder of them, boasting to their schoolfriends of this exciting adventure in urban farming.

As Christmas approached and my grandfather was preparing to 'do the deed', my grandmother went to feed the chickens and to her surprise discovered an egg in the makeshift coop. It didn't take her long to realize that an egg a day for the foreseeable future would provide considerably more food than a single roast dinner – so the chickens were saved . . .

Consequently, that Christmas, instead of the steaming roast bird that they had dreamed of, the family ate a humble pie of tinned corned beef with a sprig of holly spiked in the top.

As it turned out, all three chicks were female, which has never been my experience, and very soon a little enterprise evolved, with local neighbours buying eggs from them with their ration books, allowing Dad's family to buy extra feed for the chickens. They ended up with twelve hens and a cockerel (to father many more) and it became quite an extensive operation for many years. When I asked him about living through the Blitz, Dad explained that, of course, they didn't take the chickens down the air-raid shelter, but there was a 'near miss' for them on one occasion when a large land mine came down by parachute in the adjoining garden. Fortunately, the mine didn't go off, the chickens survived the shock, and my dad is alive and well to tell the tale today.

Keeping chickens is a delightful hobby and our feathered friends can keep us entertained hour upon hour. They are also easy livestock to keep in urban areas. When we were debating the best options for us in terms of protein from the garden, the children voted for sausages and burgers, and I had to explain that our garden wasn't big enough for pigs, let alone a cow! Andy meanwhile wasn't convinced that the broad beans would offer enough protein to fill him up properly. I'd been studying some new designer

LEFT Our Cream Legbar hen with a Pekin and Wyandotte bantam.

chicken coops called Eglus, made by a company called Omlet, and rather fancied a couple of hens in the back garden. I thought it was a great idea – the children could understand where their food came from and the animals didn't need to die for our dinner, plus they would create loads of manure for compost that we could use in growing more crops.

Andy was not convinced, so I continued to plead and it eventually became a family joke. Then one day during the school holidays he asked what my plans were for the week. When I explained that I was taking the children out on day trips, he asked me to stay in for a delivery. My retort was, 'Stay in for your own delivery!' And then he told me it was coming from Omlet . . .

Apparently, Andy had seen the Eglus displayed at a design exhibition and agreed that they were rather cool, so he'd asked friends and family to contribute towards one as a birthday gift for me. I couldn't believe that he'd conceded. But now that it was confirmed, I realized that I knew nothing about keeping chickens. I'd only ever seen them from a distance and certainly had never touched one. I mean, you don't usually touch birds, do you? They are either flying in the sky or perched high in a tree . . .

Once the Eglu coop was safely installed in the garden, the Omlet rep invited me to choose two of the hens from the crate in his van. I decided on a black one with beautiful iridescent plumes and a pure white one – and named them Pepsi and Shirley, after the backing singers of pop group Wham!

I was captivated. I could sit for ages just watching them scratching the lawn pecking for bugs. As for hugging the hens, I couldn't have imagined how delightful that would be. And here we are many years later with an ever-increasing flock of them.

THE JOYS OF KEEPING CHICKENS

For us, the chickens are food producers, and deepen our understanding of where our food comes from. They bring life to the garden and help us engage with nature. Others have chosen chickens as outside pets; this means no pet dust in the house, which can help those with allergies.

We've found them to be an inspiration to all who meet them, whether that's the children on education trips or our visiting friends and neighbours. In the UK, as long as your flock is less than fifty birds, you don't need to register them; you just ask your landowner for permission to keep them. Their output is invaluable, both the eggs for eating and the manure for composting (and there's a *lot* of chicken manure if you keep a lot of hens!).

Chickens come in a wide range of shapes, sizes, colours and feather patterns. There are pure breeds, where mum and dad are both the same breed, and hybrids, which are deliberately bred to keep all the best qualities – for example, lots of eggs and good all-round hardiness – with fewer of the traits that come with pure breeds, like long periods of broodiness (see page 121).

The bigger pure breeds are called large fowl and the smallest breeds are classified as bantams. As you'd expect, the smaller the chicken, the smaller the eggs they lay.

ABOVE LEFT An Omlet Eglu with a fox-resistant run.
ABOVE The Eggport gives you direct access to the nest box.

ABOVE Barbara, our Gold Partridge Brahma, one of the biggest breeds in the UK.

On our chicken-keeping courses I'm regularly asked about rehoming commercial hens, which I think is a great idea; but I do think, due to specific health and rehabilitation challenges, this is easier to manage if you already have experience of looking after chickens. Others say that they want to start with baby chicks as they are so cute – which they are, but baby chicks can't live outside until they are fully feathered, so you may need to keep them in your own home for weeks or months until they are less vulnerable. I always say that it's easier to start with hybrid hens as they are reliable, definitely girls, and when you buy them at 'point of lay' will be producing your breakfast eggs in no time at all.

We only keep hens (female) here at Hen Corner, as cockerels (male) are not only noisy – think *cock-a-doodle-doo* an hour before dusk and dawn – but can also be aggressive. Above all, the boys don't lay any eggs!

THE RESPONSIBILITIES OF KEEPING CHICKENS

As with any living thing that we are responsible for, we need to protect and provide for them.

PROTECT

- We protect from predators with secure housing. The Omlet Eglu has a fox-resistant run and they've kept our hens safe for 15 years so far.
- A regular routine can protect from parasites. We feed them VermX pellets one week a month for gut health and protection against worms. Regular cleaning of the coop reduces red mite and dust baths reduce lice.
- Extreme weather is something else chickens need protection from. Imagine their feathers are like a thick down jacket: it'll keep them warm, but not necessarily dry, and they can't take it off in the summer. So, shelter from rain and shade from hot sunshine will help them stay dry and cool.

PROVIDE

- Food, they only need layers pellets with some mixed corn as a treat. In the UK it's illegal to feed them kitchen scraps. I ensure that there's a constant supply of feed as I'm confident that they don't overeat.
- Water, clean and fresh as we like to drink it. They need a lot to help with digestion, egg production and temperature regulation, especially in hot weather.
- A clean, secure coop and run, with a private nest box for laying eggs and shelter from the sun and rain.

DAILY

The daily responsibility is simply ensuring that they have enough food and water and that you're collecting the eggs.

WEEKLY

Our super Eglu houses are very easy to keep clean; we just slip out the droppings tray, tip the contents into the compost bin and reline with old newspaper before sliding back in place. The nest-box straw can also be composted and then replaced with fresh bedding.

SEASONALLY

We usually deep clean the coops three times a year: before winter, after winter and midsummer. The plastic Eglus are easy to take apart, allowing us to jet wash them thoroughly and dry them in the sunshine; we then rake out the base of the run, sprinkling over ground sanitizing powder and topping with fresh woodchip.

URBAN HENS – KEEPING CHICKENS IN LONDON

While the daily needs and general requirements of chickens are the same in both urban and rural environments, there are a few things to bear in mind if, like us, you live in a town or city.

Space
We probably have less space in an urban environment. Check out coops that have run space under the chicken house because this maximizes the chicken's living space on a smaller footprint of land. Chicken feed usually comes in 20kg/44lb sacks; to save valuable shed space, we store ours in a metal dustbin that is both waterproof and vermin-proof.

Free-range opportunities
With less space, the chickens can't wander too far away if you allow them to free range – though do watch out for precious flower beds and veg patches. If you like to eat it, so will they.

Accessibility of feed and equipment
We've found over the years that chicken-keeping has become increasingly popular and we can now buy most of what we need in a local pet shop. Do check where you can buy feed from in advance as you don't want to run out when the shops are closed or while waiting for a delivery.

Dealing with waste
The main waste is from the droppings tray and nest box, which is all compostable. Make sure you have a big enough compost bin or check to see if it can be added to your council garden-waste collection.

Neighbours

We've always found our neighbours to be particularly fond of the chickens and the occasional gift of a box of fresh eggs is the perfect way to share the love.

Predators

Here in London, urban foxes are absolutely brazen, strolling around during the day in total confidence as they have no predators themselves. Because of this, we only allow our hens to free range when we can supervise them, ready to chase away any fox that might come prowling for an easy dinner.

Vets

Over the years, I've discovered that urban vets have limited experience of treating chickens. As chickens try to conceal any illness or weakness for fear of being bullied, by the time we know that they are unwell it's often too late for the vet.

Schools

Keeping chickens in city schools is a great way to allow a whole community to share in the experience, especially when the pupils may not have gardens of their own.

Allotments and community gardens

I've supported several groups as they started keeping chickens together in communal spaces. This is a fabulous way of doing things, especially when chickens are kept alongside growing fruit and vegetables. There are urban farms and co-operatives all over our city.

OPPOSITE A cooling dustbath on a hot day.
BELOW Eggs of every colour.
OVERLEAF Taking time out with 'the girls'.

THE POULTRY-KEEPER'S YEAR

SPRING

The start of the season of promise: with our hens back in lay, this means lots of eggs. A hybrid hen can lay up to 350 a year, nearly one every day, so the more chickens you have the bigger the egg basket you will need.

By late spring some of the hens – usually the pure breeds, in our experience – go broody. This means that they've laid enough eggs over recent weeks and they've decided to hatch them. Of course, I tell them that nothing will happen, that there are definitely no chicks inside and, ahem, the silly hen hasn't even got a boyfriend, but they still sit and wait. So committed is the broody hen that she denies herself food and water for most of the day and plucks the feathers from her breast to reduce insulation from her body warmth and humidity. She is so protective of her imaginary offspring that she growls at anyone, person or hen, that approaches her. We remove the eggs from under her and we remove her from the nest box, but she will still find a way back to her sacred spot and can stay, waiting, for months – long past the usual twenty-one days of incubation – until she eventually abandons her efforts, accepting that life is more rewarding out in the sunshine.

SUMMER

We need to remember that chickens, with their wonderful fluffy feathers, can't simply slip off their coats, so providing shade is crucial. Longer days mean less poop in the coop, but the warmth of the sun and humidity of feathered bodies is the ideal environment for mites and lice. Keep an eye on the chickens. If they have access to a dust bath, they should be able to keep lice under control. Watch for scales lifting on their legs and feet; if you spot this telltale sign of scaly leg mite, treat it with a spray. Regular cleaning of the coop will help to keep red mites at bay.

HOLIDAYS

We've found that neighbours are usually more than happy to feed and water the hens, collecting the eggs as their reward. However, asking friends to scoop poop for us probably goes beyond the call of usual friendly help! In the summer, it should be okay to leave the droppings tray for a couple of weeks while you go on holiday; just clean it out before you go and on your return.

AUTUMN

Now that the days are cooling down, the hens begin to grow new feathers in preparation for winter. Old feathers drop out in a moult, which can vary from a discreet one or two here and there, through very scruffy and on to the 'oven-ready' look. I've been caught out on several occasions, finding a pile of feathers and fearing a fox feast only to discover that beautiful new plumage is on the way.

As the days get shorter, the hens are in the house for longer as they naturally retire at dusk. This means more poo to clean out. This is an ideal time to do a deep clean in preparation for the colder months.

WINTER

Remember that birds are outside creatures and chickens are particularly hardy. They can cope with cold temperatures really well, as long as they have sufficient food; they will grow extra feathers and huddle together to keep warm at night.

Something to be careful of in winter is that when it's really cold their water drinkers can freeze, leaving them very thirsty. Simply empty them each evening and top up in the morning. Recent years have found UK government restrictions for all poultry-keepers to protect their birds from avian flu, which can be transmitted by wild birds – particularly those that migrate here in the winter. Follow local guidance as it's announced and your hens will stay safe and sound.

BEES

I f Andy was hesitant about keeping chickens, he was putting his foot down now when I tested out the idea of keeping bees in the garden. 'It'll be wonderful for pollination,' I said, 'Getting more apples on the trees and all those raspberries that you like, not to mention the beautiful honey . . . ' 'No!' he said. 'We haven't just moved to this nice new house to have bees flying all over us when we're trying to eat a barbecue. I want to lie in the hammock without bees bothering me. We've got a lovely garden for us to enjoy; this is not a farm. Plus I don't even like honey.'

Mmm, I could see that this was going to be another process of persuasion. I completely understood that keeping bees was not something to rush into – you don't just pop down to the local pet shop and ask for a box of 40,000 small stinging things! – but surely it could be possible?

Then a Christmas present from Andy gave me a glimmer of hope: a beekeeping book and a day's course at River Cottage in Dorset, albeit with a note saying, 'This does NOT mean that we are getting bees.' This was a step forward and I was very excited.

The beekeeping course was fascinating. We started with classroom learning – they even had live bees behind glass in the room – then went out to visit two apiaries, one in local woods and one on the farm. I'd followed the guidelines on what to wear to reduce the risk of getting stung and now, wearing a smock and veil, I waited patiently for my turn to hold a frame of live bees. I held up the frame with the light behind me, allowing me to watch the nurse bees fastidiously care for the developing larvae, spotting corners of sealed honey stored safely nearby and marvelling at the multiple colours of packed pollen. I knew that I could do this.

At the end of the day, I asked the course tutor, an experienced beekeeper, for any final tips for keeping bees in the back garden. 'Yes,' he said. 'Don't do it!'

I was perplexed: surely this course was to encourage us to go on and keep our own bees. Why on earth should I not keep bees in my back garden?

He then went on to suggest that I ask a local farmer for a corner of his field, not really an option in our part of densely populated urban London!

Fortunately, it wasn't long before I discovered a local beekeeping association along with a friendly beekeeper who assured me that lots of people keep bees in their back gardens and offered to help me on my journey towards Plan Bee. The following year brought more training, family visits to apiaries and checking out a back garden with happy hives. Andy still wasn't convinced. Just as I began to feel we were reaching the

LEFT Products of the hive.

end of the road with this ambition, the friendly beekeeper suggested that I borrow a colony of bees: if we liked them we could keep them and if it didn't work out we could return them. It sounded perfect.

While not as ecstatic as me, Andy could see the benefits of this arrangement. He was also confident that he would be proved right and that the bees would be returned to their previous owner.

Of course, the bees stayed – and ten years on, all of the bees in the garden at Hen Corner are the offspring of that original colony we borrowed. Plan Bee worked out well in the end.

A BRIEF HISTORY

Honeybees are just one of hundreds of species of bee here in the UK. They live in the biggest families, known as colonies, and make more honey than they need. This allows us to take some of their extra resources for ourselves. Well before the cultivation of sugar, honey was the sweetest food available. An 8,000-year-old cave painting has been found in Spain depicting a brave individual climbing high to pull a comb of honey from wild bees. Collecting it involved a real risk versus reward, but it was such a treasured possession. Ancient Egyptians would bury honeycomb along with their dead kings in the pyramids; thanks to its eternal properties – it doesn't perish or expire – it was perfect for the long journey into the afterlife.

The tradition of keeping bees for honey, in a hive such as a traditional straw skep, began a few thousand years ago. While skeps were easy to make and provided a perfect home for the bees, they didn't allow the beekeeper easy access to check on them. When trying to collect the honey, not only were you likely to get stung but the delicate wax combs could also fall and damage the bees' well-kept home and young offspring.

Around 1850, Rev. Lorenzo Langstroth, of Philadelphia, USA, discovered 'bee space': the distance allowed between honeycombs constructed by bees in wild nests. This exact space permits bees to walk across a sealed comb back-to-back against the bees working on the next comb. If the space is too tight, only one side of the comb will be utilized; if the space is wider, more wax comb gets built in the gap. It was this discovery that inspired Langstroth to design a beehive with moveable frames, evenly spaced, holding individual honeycombs. This would allow a beekeeper to inspect a colony of bees frame by frame, comb by comb, without much disturbance to the bees.

Langstroth's design revolutionized the domestication of honeybees, improving the bees' health and also our harvests.

A MODERN-DAY HIVE

While there are many types of beehive in use today, most consist of a brood box that holds the frames of the brood nest. It's a bit like a filing cabinet: you can lift out the frames and 'read' what's happening in the hive. These brood boxes have an entrance for the flying bees to come and go and are set on a stand to lift it from the ground and keep it at a height easy to work with.

Directly on top of the frames sits a closefitting board to prevent the bees building wild comb in a gap. Finally, there is a roof on top to keep everything warm and dry. When the brood box gets full of bees working on every frame, with most cells filled with brood or food, we can add another box of frames to give them more space. If we position a grid, called a queen excluder, between the bottom box with the nest of babies and the upper box, the queen can't climb up into the new box – so the bees will just use this space to store honey. As honey is heavier than eggs and larvae, we use shallower frames; each of which can hold up to 1.4kg/3lb of honey, so a box of twelve frames, full at harvest time, is very heavy to lift!

THE ROLE OF THE BEEKEEPER

As beekeepers, we receive valuable benefits from these little insects in their role as pollinators and honey producers. But caring for them involves many responsibilities that we need to take seriously – for the sake of their health and well-being and for the wider community that we live in, especially in towns and cities.

ABOVE Omlet Beehaus.

WHO IS IN THE HIVE?

A honeybee colony includes three types of bee:

A QUEEN BEE

A queen bee is the mother of the family, and lays eggs all day long (up to 2,000 a day in the summer). She is larger than the other bees and can live up to five years. She will only leave the hive for her mating flight, just days after emerging from her cocoon, and later if her daughters encourage her to swarm to a new home, which is a natural part of honeybee reproduction. To help spot the queen among up to 60,000 other bees, many beekeepers mark them with a spot of coloured paint. This also helps identify how old she is. Queen-bee marking uses five different colours applied in sequence, each colour linked to a year: white for years ending in 1 or 6, yellow for years ending in 2 or 7 and so on. The mnemonic 'Will You Raise Good Bees' is used to remember the sequence. So, for example, new queens mated in 2023 and 2028 will be marked red.

DRONES

Drones are the male bees. They have a lifespan of just a few weeks and are present only in large, healthy colonies during the warmer months. They don't work; they don't sting; they don't contribute to the colony at all. They are fed, cleaned and cared for by their sisters. They are simply tolerated (when there are sufficient resources) for the greater good: to potentially mate with an unrelated colony's virgin queen and share the good genes around. A drone lucky enough to fulfil this sole purpose will die in the process – leaving his genitals inside the queen, glowing ultraviolet light to attract other drones. Drones who try and fail this quest and return to the hive may be evicted in late summer when they become an unnecessary drain on resources during winter preparations. Here they'll have their wings nibbled off by their sisters and then be ushered outside to die.

WORKER BEES

Worker bees are the girls and form the majority of the hive population. During the winter, when there is little food outside of the hive and the priority is protecting the queen through the colder weather, the girls live around six months, from September through to March. In the summer, their intensive work wears them out by the age of just six weeks. They spend the first four fulfilling important roles inside the hive, such as nurturing the young, making wax for comb, ripening honey and general housework/cleaning duties. The remaining fortnight permits foraging outside of the hive, within a 2.4km/1.5-mile radius, looking for nectar, pollen, water and propolis – all vital for the hive's sustenance and safety.

- **Nectar** is an energy-giving carbohydrate. It can be stored as honey when the water content is evaporated to less than 20 per cent. Once it's 'ripe', the bees seal the honey cell with a wax capping – like screwing a lid on a jar.
- **Pollen** is protein, perfect for developing larvae. It collects on a bee's body while she has her head in a flower sucking up nectar; it then dusts off on to the other flowers she attends to, starting the magic of pollination. She uses her middle legs to brush remaining pollen into pollen presses on her hind legs, allowing her to carry it home to the hive for safe storage.
- **Water** helps with temperature control, which is important when trying to maintain a steady 36°C/96.8°F inside the hive with up to 60,000 bees focused on their specific tasks.
- **Propolis** is collected from certain tree saps and sticky buds. This fragrant, resin-like substance has multiple uses within the hive. With its natural antibacterial properties it can be used to polish the empty cells before the queen lays her eggs, to fill gaps and seal areas of the hive and, in extreme cases, even to mummify an intruder such as mice – which bees can kill by smothering but are too large for them to remove from the hive.

Throughout the main beekeeping season, April to July, the nest of honeybees is at its largest and this is when we need to be most vigilant, performing weekly inspections to monitor the 3,500 cells on each side of 11 brood frames in every hive. I don't expect the bees look forward to our inspections, when we prise the top board away from the frames and allow dazzling sunlight into the pitch-black hive. But, if we are careful, we can foster a gentle, respectful relationship that doesn't disturb the bees too much and minimizes their instinct to defend the hive by stinging us, the intruder.

SMOKING THE HIVE

Using smoke when we open the hive can make inspections easier. Contrary to popular belief, the smoke doesn't actually calm the bees – smoke puffed in my face would do the exact opposite – but there are three main reasons to use smoke and they all make sense:

- **The bees move away from the smoke.** Just like you or I would. This is helpful when we want the bees to move out of the way so that we can lift out a frame without squashing them, or get a good view of what's below them.
- **The bees think the hive is on fire.** And as wood, wax and honey are each very flammable, they prepare to leave. They don't know where they will go, because this is not a planned swarm; and they don't know how long it will be until they find a suitable new home. They *do* know that once settled they'll need to build new wax comb before the queen can start laying again – and making wax requires a full tummy of honey. Unlike us, bees don't rush to vacate the building during a fire drill. Instead they wait for the command to flee and in the meantime gorge themselves on honey, like a huge packed lunch for the journey. Of course, the hive is not actually on fire. However, in

ABOVE Bees working on the honey frames.

this post-lunch, docile mood they are less interested in the beekeeper. I like to imagine the conversations they might have if bees could talk. 'Wow, I'm stuffed – a bit tired now actually . . . ' 'Yeah, after that big meal, I could do with a snooze . . . '

- **The smoke masks their messages.** Bees communicate with pheromones, which are recognized a bit like smells. There are many pheromones wafting around inside the hive. The queen's has a message like, 'I am your mummy and I love you'; when workers sense this they naturally feel secure and comfortable. Guard bees at the entrance of the hive waft the 'home' pheromone: 'If your name's not down, you're not coming in.' This tells wasps, bumbles and honeybees from other colonies they're not welcome. Having a beekeeper poke around the hive is a bit like having a giant remove your bedroom ceiling and turn your bed upside down for a better look. The bees' natural response is fear and soon they will all be pumping out the associated pheromone, 'alarm'. Fortunately, though, the smell of smoke is stronger than the smell of fear.

The weekly inspection, ideally on a warm sunny day, is an opportunity to check on their brood, food and mood, looking out for any health issues and ensuring they have enough space. We also want to discover their secrets, especially if they are planning to swarm.

BROOD

The collective term for the babies in the hive is brood and it can be found in three stages:

- **Eggs** are the shape of a grain of rice but smaller than a fine thread of cotton. They hatch after three days, releasing tiny larvae (the bee equivalent of a caterpillar).
- **Larvae** are fed by their older sisters with a nutrient-rich diet of royal jelly, secreted from glands in the heads of young nurse bees, followed by bee bread made from pollen and nectar.
- **Sealed brood.** Six days after the larvae hatch (or five for queens and seven for drones), they are sealed into cells with a capping of wax mixed with pollen. This allows a cocoon to be spun to protect the larvae during metamorphosis into adult bees.

The whole brood cycle takes twenty-one days for worker bees, sixteen for queens and twenty-four for drones. When we inspect the hive, we hope to see healthy brood in all three stages, distributed in an even pattern. The nest of workers tending to the developing brood is usually rugby-ball shaped, with a large round of brood on the hive's middle frames and smaller rounds towards the outer edges. In the height of summer, the nest will almost fill the brood box, but in the winter, with less forage available, it will shrink down to maybe the size of your fist.

LEFT Smoking the hive.

FOOD

Ideally, our honeybees would be foraging among a wide variety of flowers, bringing back a diverse mix of nectar and a rainbow of pollen. Surely the countryside would be perfect, with its farms and fields already planted up with food for us? Unfortunately, farmers are under pressure to produce the largest harvests possible, so extensive fields are filled with a single crop, such as oilseed rape, which renders a high yield of valuable product. This offers a delectable feast for the bees while in flower, usually during the four weeks of May, but then leaves a derelict buffet table with hardly any food for the subsequent teams of foraging bees.

Imagine the excitement of the bees on discovering a huge yellow field of food – surely the message would go back to the hive, along with a waggle dance, that more bees are needed to bring home the bounty. The queen might be encouraged to lay more eggs so an army of foragers can be raised. But let's just do the maths. It takes three weeks from egg

ABOVE Honeybee foraging for nectar.

to adult bee and then there's four weeks of housework. Therefore, the youngest foraging bees would be available seven weeks after the message was sent. Will that crop still be flowering seven weeks later? Probably not. This leaves all the new bees hunting around the hedgerows and borders for scraps of bramble flowers, and what's there probably won't feed 2,000 new mouths.

As beekeepers we must ensure that there is enough food stored in the hive for the growing colony and recognize that there are ebbs and flows in nectar. For example, there can be a break in available forage between the spring flowers and blossom and the later summer flowers. I've also found that a very dry spring and summer may provide a vast range of flowers, but if there's insufficient moisture in the ground those flowers simply can't yield much nectar. If food stores become depleted, especially in a new, smaller colony, we can offer them sugar syrup in a feeder within the hive to support them until more forage becomes available.

ABOVE Ripening honey in the comb.

MOOD

It's been said that a hive humming at 'middle C' is a happy hive; and while I don't take a tuning fork with me on my inspections, I can tell quite soon what mood my bees are in. Usually, if all is well and healthy, the kids are happy and there's enough food to go round, your colony will have little to worry about and will simply focus on the tasks at hand. However, if their behaviour is erratic, or they seem stressed by your presence and generally rather defensive, something might be wrong.

Things that can affect a colony's mood:
- Hunger – see previous page
- Missing queen – see next page
- Cold – we can provide extra insulation in the winter
- Air pressure – they can sense when the weather is about to change
- Feeling threatened – worker bees will surrender their own lives by using their barbed sting to defend their precious brood or food
- Obstruction – we are simply standing in their flight path preventing them access to their home.

HEALTH

Many things can affect the health of a colony. The most common problem is varroa mites, which latch on to the backs of adult bees and slip into brood cells just before they are sealed to feed on the developing larvae. Keeping it under control is a constant battle and we treat each colony for varroa in different ways at different stages of the year. The parasitic disease nosema is another problem, which can be indicated by streaks of excrement on the outside of the hive or changes in brood appearance. I find that as we become familiar with a healthy colony, and look after them well, we become better equipped to notice problems. When something looks wrong, we can seek support and advice from either a more experienced beekeeper or our local bee inspector.

SPACE

I like to make sure that there is always a frame or two either side of the brood nest, allowing the bees extra space for expansion. When the brood box becomes full, I'll put super boxes on top so there's no restriction as they store their honey. I've learned that bees are less likely to plan to divide and swarm when they don't feel tight in their hive – but, as a beekeeping mantra reminds us, 'The bees don't read the same books as us.'

SWARMING

Long, protruding queen cells may be seen hanging down vertically within the hive for several reasons. In my experience, it's usually one of the following:

ABOVE A swarm of bees

- The queen is old and less fertile and the workers plan to replace her.
- The queen is missing (maybe I squashed her!) and the workers plan to replace her.
- The queen is about to swarm with up to half the colony to a new home already determined by scout bees. The remaining bees will need a queen, so the workers plan to replace her.

It's important for us beekeepers to try to understand what the bees are planning. In preparation of rearing a new queen, they will sculpt a larger wax cell to hold her long, elegant cocoon. They will also ensure that she is only fed precious royal jelly at the larvae stage and seal her cell securely just eight days after the egg was laid. If they are going to swarm it will usually be on the day the new queen cell is sealed, so it's important to inspect regularly within eight days. When they plan to swarm, they'll usually have a few new queens on the go as insurance, which can potentially trigger multiple caste-swarms that deplete the population of the original colony. So, when we see queen cells, we need to 'read' what the bees are planning and try to keep a step ahead, responding accordingly.

COUNTRY VERSUS CITY

In a rural area, swarming would be a very natural form of colony reproduction, with the scout bees looking for appropriate space and shelter in a tree hollow, under the eaves of an old barn or in a disused oil drum and so on. In the city, a swarm might choose a chimney stack, a cavity wall, a roof void or some other inconvenient place. Bees are less likely to sting when swarming, as they have no brood or food to defend, but in a densely populated area, having 30,000–40,000 bees flying around you can be very overwhelming. And if they build a home within your home, they can be difficult and costly to remove.

A big advantage for bees in towns and cities is the wide variety of food available. Parks and gardens are often planted for year-round flowers, allotments contain multiple crops that need pollinating, and blackberries grow wild – so urban forage can be much more diverse than rural forage and is often available for longer.

A bonus for urban beekeepers, particularly here in London, is that there are so many of us relatively close together. This is great for support, mentoring, sharing equipment and general encouragement. However, there's a well-known saying about seeking advice: 'Ask two beekeepers a question and get five different answers . . .'

THE BEEKEEPER'S YEAR

SPRING

Watch to see if the bees are flying on a sunny day. A good indicator that there is new brood to feed is if they are carrying in pollen. Start your weekly inspections.

AUTUMN

Prepare your bees for winter with varroa treatments and feeding. Enter the National Honey Show and catch up with great talks, lectures and workshops.

SUMMER

Continue weekly inspections, adding boxes for honey and watching out for swarming. Harvest honey at the end of summer.

WINTER

Check the hives for food stores and treat again for varroa. Make up new frames ready to replace old ones in spring.

LARGER LIVESTOCK

With the care of chickens and bees comfortably in our daily routine, it's easy to wonder what other animals could be kept for food.

I remember the day that I invited Steve Paynter and his son, Jono, round to meet the bees. Steve is the vicar of a West London church and we've worked together now and then. I'd heard that they were fellow chicken-keepers and also reared pigs in the vicarage garden. I didn't actually believe this, as I was sure that it would have come up in conversation before – but after checking it out and securing an invitation to see, I discovered that they had a couple of chicken coops and a substantial enclosure for two Gloucester Old Spots. I was very impressed and slightly jealous, especially as we'd previously bought a quarter of a pig from a friend of a friend who rears them in Hampshire and it was the best pork ever. And these guys are doing it just down the road! Andy won't let me have a goat (even though I've thought it all through and plan to convert our small garage into a goat shed and take it to the park each day along with the dog walkers . . .) so a pig is definitely out of the question. What a shame. Steve had been thinking about bees for a while and was keen to see how it could work in his garden and, as it turned out, we swapped a spare colony of our bees for a couple of their home-hatched hens and both parties were very pleased with the deal.

Unlike with chickens and bees, which feed us year on year with their eggs and honey, rearing larger livestock for meat means a shorter life for the animal, and meat every autumn means new animals every spring.

When I asked Steve and Jono if they were planning to keep pigs again in the new year, once these two were slaughtered, Jono said that he wanted to rear some Saddlebacks. I suggested that I could pay for one and cover the cost of its upkeep if he'd be willing to look after it for me. He said yes! So, I had my pig and ate it after all.

While we enjoyed the pig share, and helped out with the feeding when they were on holiday, I was rather pleased that it was kept in the vicarage garden and not ours. But I still nurture the dream of keeping a goat for milking, allowing me to make lovely fresh cheese. However, I completely appreciate that it's one thing getting neighbours to feed the cat and collect the eggs while we are on holiday, but twice-daily hand milking a cheeky nanny goat may be asking a bit too much!

Elsa, a beekeeping friend, seems to have trumped us all because, alongside her bees, she has chickens at home and keeps pigs, goats and a cow on an old farm that is being run by volunteers not far from a West London underground station. It's always a delight

LEFT Elsa's Tamworth pigs in the woodland.

to pop up to the farm to see her pigs. She had Tamworths and Gloucester Old Spot crosses last time I visited.

While Elsa has lots of space to keep them free range at the farm, a couple of pigs (they are very sociable and like to share with a buddy) only need 100sq. m/1,100sq. ft. Within this space they need shelter, such as a pig ark, for sleeping, and outdoor space for feeding, exercise and foraging. If a pig doesn't have space to run and root around they can become bored and fidgety, which could result in a bit of antisocial behaviour.

Unlike chickens (if you keep less than fifty) and bees, there are important regulations that you need to comply with for larger animals, which are in place for the well-being of both the animals and those eventually eating them.

In the UK, the legal requirements are based around the following:

- **Registering your land** to get a county parish holding number, confirming where the animals will be kept.
- **Registering your herd**, even if it's just a couple for pets, giving you a 'herd mark' for your individual animals.
- **Reporting the movement of animals**, when they arrive and leave your plot, to prevent and control disease.
- **Record keeping** of feed purchased, medication used, any changes to land or herd.
- **Animal welfare**, ensuring your animals have all they need, including the correct food, unlimited drinking water, open space, secure fencing and so on.

WHAT GOES INTO A PIG?

Pigs will eat almost anything, so it's important to ensure that their main feed is a balanced diet. As with chickens, it's illegal to feed food scraps to pigs as traces of meat can pass disease into the food chain. The rule of thumb is to avoid feeding livestock any food that has been through either a commercial or domestic kitchen. Fresh veggies straight from the allotment are fine, as are morsels that they find as they forage; but for their main diet, pellets of complete feed ensure that all their nutritional needs are met.

WHAT COMES OUT OF A PIG?

Pigs, while they love to roll around in cooling mud, are very clean animals when it comes to toileting and will usually return to the same corner of their pen to use as a latrine, which makes cleaning out so much easier. Their dung can be cleared away regularly and added to compost to rot down nicely.

DAY-TO-DAY CARE

While it's wonderful to simply watch them wallow, pigs require a couple of visits a day to check that their water is clean and to feed them. Unlike chickens, they don't regulate their eating and will overeat if they have access to more than they need. So strict rations, for breakfast and supper, are the way forward.

While you're with them, a quick visual check can confirm that they look happy and healthy and have no obvious injuries. As with all animals, being familiar with how they are when they are fit and well helps us to recognize when something is not quite right. Also, have a quick look around their pen. Are they trying to break free? Are there weak points in the fencing? Have they dug up something dangerous, like an old glass bottle or broken china?

In wet weather, the straw bedding in their ark might need changing. If you are keeping pigs all year round, you might want to relocate their pen to avoid damage or contamination of the ground.

Finally, don't forget to register with a vet. If you live in a town or city they may well have to drive in from the countryside, but a rural vet should have more experience with larger livestock.

SAYING GOODBYE

If you are growing 'weaners' for meat, it's usual to buy them around eight weeks old and send them to the abattoir around six months old.

As Elsa says, 'It's never easy to send them off, but I know they've had the best life ever . . . running, playing, digging and wallowing in their mud bath which they love.'

So, if you can bear to say goodbye, it'll be bacon for breakfast and sausages for supper.

ABOVE Feeding time for the piggies.

CELEBRATE

 We mark the seasons and the years.
We tell people that we love them and we
look forward to the blessings ahead . . .

Living with the seasons and producing your own food means that there are many celebrations, big and small, throughout the year.

The first spear of asparagus, a new chick hatching, collecting a swarm of bees, digging up potatoes, weighing the biggest pumpkin, plucking ripe figs . . . There are so many things to be thankful for.

Gathering family and friends to celebrate throughout the year, building faith and sharing joy, are significant events for making memories and expressing gratitude.

As we look around us, yes, there are many things that could be better, people who could be nicer, loved ones who could be healthier (or still with us). The environment we are asked to care for is often not treated with the respect it deserves. Nations fight, children starve and those in authority don't always make the decisions that we'd want them to. In all these things, we do what we can to help and pray for the situations that we can't change ourselves.

Amid this, we mark the seasons and the years. We tell people that we love them and we look forward to the blessings ahead.

Celebration is good for the soul.

VALENTINE'S DAY

While Valentine's Day isn't celebrated by all, it is a very special day in our house – though not for the obvious reasons.

Yes, Andy and I exchange cards and small gifts over breakfast, but as soon as we've opened them the focus of the day is flipped, because 14 February is Andy's birthday and the rest of the day is centred around him. Well, almost.

We are Hen Corner, after all, and our flock of ladies does need a mention, as Valentine's Day is usually associated with them coming back into lay after their winter break. It makes perfect sense if you think about it. Chickens lay eggs primarily for their own reproduction, not for our breakfasts, and a cold, dark winter is not the ideal time for nurturing baby chicks. While hybrid hens, who have been bred for maximum egg production, may well lay regularly throughout their first winter, older hens and pure breeds will have a natural break and will begin again as the days start to get longer. The longer days, as we approach the spring equinox, allow more sunlight to hit the pituitary gland, entering through the eyes of the hen, which triggers hormones to start egg production. So, after breakfast on Valentine's Day, I open the nest boxes with anticipation, hoping to see a nice clutch of eggs to make Andy's birthday cake with.

We tend not to go out for dinner on his birthday, as the restaurants are often heavily priced and themed for lovers, so we've developed a new tradition of eating a feast at home, serving Andy's favourite food. The main course is goose confit, which I start to make on Christmas Eve when I'm preparing the goose for the next day. I remove the legs and wings to shorten the roasting time of the bird and prepare them in a traditional French preserving technique called confitting. The meat then keeps perfectly for months, sealed in its own seasoned fat. This delicacy is served with vegetables from the allotment – potatoes cooked in cream as pommes dauphinoise and steamed broccoli. Dessert will be relying on eggs again, it's usually crème brûlée.

One year, Andy bought me the perfect Valentine gift, a 'grow your own' mistletoe kit. Mistletoe is often associated with Christmas and with kissing – making it perfect for Valentine's too! To cultivate your own mistletoe, you need the fresh white berries that, in the UK, aren't usually mature until February. Mistletoe is a parasitic plant that feeds off an established host plant; it particularly favours apple trees, among others, and I'd longed to harvest my own from the garden for many years. I love searching for clusters of mistletoe, hanging like chandeliers at the top of tall trees naked of their leaves in winter, but as new plants are usually propagated by hungry birds who eat the berries and

LEFT Valentine's Day and the hens are back in lay.

poop out the seeds when they are roosting high in trees, their evergreen bouquets are often well out of reach.

After several winter walks, spying for any berries left on the trees and realizing that even if we could see them, we couldn't pick them, Andy bought me the kit. Beautiful fresh berries were packaged up with labels and full instructions. We needed to pop the seeds out of the berries and stick them to branches of the chosen trees. Fortunately, nature provides an adhesive gel surrounding each seed, which helps it stay in place on the branch; and to help the process along we also wrapped a little 'cage' of wire mesh over each propagation point to protect the precious seeds from birds and squirrels alike.

Now, several years later, we have numerous bunches of mistletoe happily growing within reach and hope to spot the festive berries for seasonal kisses later in the year.

ABOVE Mistletoe in the springtime.

CRÈME BRÛLÉE

This top treat of a dessert can be made ahead and kept in the fridge for up to three days before serving.

SERVES 6

INGREDIENTS

500ml/16fl oz double cream

6 free-range egg yolks

100g/4oz caster sugar

1 tbsp vanilla bean paste

4 tbsp demerara sugar for the topping

METHOD

1. Preheat the oven at 150°C/300°F.
2. Warm the cream in a saucepan until it starts to simmer, remove from heat.
3. In a large bowl, whisk together the egg yolks, caster sugar and vanilla bean paste until pale, thick and smooth.
4. Gently pour the warm cream into the egg mixture whilst slowly whisking it in.
5. Pour the mixture into six small ramekins or one larger ovenproof dish.
6. Place the ramekins/dish into a large roasting tin and top up with boiling water to two-thirds of the way up the dishes.
7. Bake in the oven for approx 40 minutes, until the desserts are nearly set with a little wobble in the centre when you tap them on the side.
8. Carefully remove the ramekins/dish from the hot water and allow to cool to room temperature before chilling in the fridge until needed.
9. To serve, sprinkle demerara sugar on the top of each dessert and carefully place under a preheated grill until the sugar melts and caramelizes.
10. Allow the sugar to cool and set to a crisp before cracking the top with a spoon to eat.

EASTER

W hen I was a child growing up in a Christian family, Easter was almost as big a celebration as Christmas. The four-day weekend, which is now seen by many as just a great opportunity for home improvement or gardening tasks, has a specific focus on each day. We start remembering that first Easter the Sunday before with palm crosses and stories of Jesus on a donkey. On Thursday evening, we may join in a Passover meal before a quiet church service of reflection on Good Friday, which was strangely followed by warm buns. Now, I've never been one to decline baked goods, but a treat, albeit with a cross on top, didn't seem a fitting companion to our sombre mood as we meditated on the crucifixion.

Saturday was always a bit odd. On one hand it was a usual Saturday, with shops open and family television shows, but there was always a nagging thought in the back of my mind: *If Jesus died yesterday and rises again tomorrow, surely this isn't just an ordinary Saturday?* All contemplation faded away by Sunday morning as the big Easter celebrations began: chocolate eggs, a lively church service, extended family for a huge feast, favourite films on telly and a later bedtime if we were quiet and Mum forgot to send us up.

Nowadays, with my own children, we have not only continued with many of the same traditions but also created new ones that dovetail nicely with life at Hen Corner. Easter is often celebrated with eggs as a symbol of new life. Jesus had a new life when he rose from the dead and Christians believe that we can have a new life, a fresh start, as we follow his teachings. So, if eggs are central to the celebrations – we've got plenty!

Since we've had chickens, Andy and the children have decorated blown eggs to display in egg cups as place settings for Easter Sunday lunch. The eggs from the wide variety of hens we keep are many different colours: white from the Leghorns, pale pink from the Orpingtons, blue from the Cream Legbars, brown from the Marans and green from the Columbines. These eggs, with decorated names, have been kept safely and the range has expanded over the years as new guests have joined us and the family has grown.

With a beautiful stash of decorated eggs, the next tradition was to make an

LEFT Real eggs for Easter.
RIGHT Decorated eggs for place settings.

Easter Tree. Using twigs of apple blossom in a vase of water, we hang the eggs like baubles; and over the weekend, the blossoms start to open in a stunning fragrant display.

I started to feel so privileged to have created a nostalgic Easter for my own family and wanted to share it more widely with local families. I wanted to create an opportunity to encourage a celebration of spring, seeing the garden in blossom with the bees pollinating the apple trees, allowing other children to collect the eggs from our chickens and to enjoy hugging the hens. *I run a bakery, so let's all have some hot cross buns – even better, let's make them all together.* Now, a few years down the line, we run Easter at Hen Corner, a popular event for local families to celebrate with us in the most tangible way.

After introductions over drinks and cupcakes, our guests join in Easter craft activities, with the adults warned that they can't join the egg hunt if they don't make a basket! Everyone, from the youngest toddler to the oldest grandfather, makes a batch of four hot cross buns to take home; we all get to cuddle a chicken and collect their eggs and then, while the buns are in the oven, there's a hunt for chocolate eggs.

This year was especially wonderful as we decided to hatch a clutch of eggs as part of our celebrations.

It all began in a Sunday School team meeting when we were discussing plans for Easter and someone nudged me and asked whether we could have some chicks. My initial response was, 'It's not that straightforward,' remembering the last time that we hatched chicks under a broody hen in the back garden and all six revealed their masculinity by crowing an hour before dawn the following Boxing Day. Sadly, we couldn't keep them as boys together are not only noisy but can be very aggressive towards each other too. But it did get me thinking.

Newly hatched chicks are lovely, tiny, fluffy and cute for the first ten days; then they start that awkward transition of adolescence, growing odd feathers in unexpected areas and eating you out of house and home. Speaking of which, if we wanted to hatch them at church so that local families could visit them in the school holidays, then we'd be using an incubator and heat lamps to mimic their mum. We'd need to parent them until they were big enough to hold their own with the rest of the flock here at Hen Corner, and this could mean them living inside our home for weeks, even months . . . *Have I really thought this through?*

If we wanted to be sure to get some new hens at the end of the project, we would probably need to hatch twelve eggs, as eight of the chicks could be boys. Twelve little chicks can happily spend their first couple of weeks in a plastic box the size of a small fish tank, but fast forward a couple of months and they'll be needing much more space.

Finally, I decided that if as many people as possible could enjoy the delight of these symbols of new life, as we reflected with thankfulness on new opportunities in our own lives, then I was happy to make this happen.

I borrowed an incubator from a friend of a friend and set the eggs to hatch on the Sunday before Easter. At church we started promoting an Easter Chick Club in the school holidays so that local children could hear the Easter story and see what was inside our real Easter eggs. Every time it was mentioned, I gulped with fear. *What if I've done it wrong? What if they don't hatch?* The eggs had been posted by a farmer friend and she wasn't sure which hens had mated, so they could all be duds.

The day before the first Chick Club, I was digging at the allotment when my children phoned to say that there was a crack on one of the eggs in the incubator. Oh my, this was really happening! When I got home, one egg was almost cracked open and we could hear other new chicks chirping from within their shells – phew, there was new life inside. The first chick that completely broke free had to wait a while to meet its friends and I started to wonder whether it was cheeping for joy, fear, hunger or loneliness. But later that evening, another hatched and when I popped down to check on them in the middle of the night, a third was hopping around.

I took the incubator to Easter Chick Club on Monday morning. The older children were excited but the younger kids didn't understand what all the fuss was about. By Tuesday, we had seven strong chicks and we allowed each child to stroke one as an adult held the chick securely. On Wednesday morning, the last day of Chick Club, after the story, we asked all the children to sit on a long step. While supervised, we allowed each child to hold a new fluffy chick and marvel at how something so small and precious could come out of a boring old egg.

The chicks stole the show at Easter at Hen Corner, and as they grew, we discovered that only two were boys, who have now gone on to new homes with breeders; we kept the five girls here to add to our flock.

It was definitely an opportunity to celebrate new life.

ABOVE Newly hatched chick.

PLANT-BASED HOT CROSS BUNS

Hot cross buns are available in the shops most of the year, but we always save them as an Easter treat . . .

We use an enriched dough for this recipe, adding coconut milk, dairy-free spread and sugar to a traditional basic bread dough. Dried fruit soaked in tea, the traditional cross on top and a sticky sugar glaze all finish them off perfectly.

MAKES 12

INGREDIENTS

For the dough
500g/1lb strong white flour
1 tsp dried yeast
1 tsp salt
50g/1¾oz caster sugar
310g/11fl oz coconut milk
(reserve the rest of the tin)
60g/2oz dairy-free spread

For the fruit and spice mix
250g/9oz mixed dried fruit
3 tsp mixed spice
1 large mug of strong Earl Grey tea

For the crosses
1 tbsp plain flour
2 tsp oil
a pinch of baking powder
1 tsp caster sugar
reserved coconut milk as required

For the sticky glaze
1 tbsp sugar mixed with
2 tbsp hot water

METHOD

1. Soak the dried fruit and spice in the hot tea to infuse while you make the dough.

2. Without shaking the tin, weigh out the coconut milk, scooping out the floating cream to use first (and saving the remainder for the cross mixture).

3. Combine with all the other dough ingredients in a large mixing bowl and work the dough for 15 minutes. Then shape it into a tight ball, place in a large bowl, cover and leave to rest for 1 hour.

4. Preheat oven to 200°C/400°F.

5. Once the dough has doubled in size, turn it out on to a lightly floured surface, lift the corners and ease into a large rectangle. Drain the mixed fruit and scatter it over the dough before folding it in, several times in different directions, until the fruit is evenly distributed.

6. Divide the dough into 12 equal portions, shape into tight buns and place on a parchment-lined baking tray. Cover and prove until doubled in size.

7. Combine the ingredients for the cross paste, using coconut milk for consistency. Pipe the paste on to the buns.

8. Bake at 200°C/400°F for approximately 12–16 minutes, until the buns are firm, risen and golden.

9. Remove from the oven and brush with warm sticky glaze.

HARVEST

Early autumn begins with wearing shorts in the sunshine and ends with a chill in the air that beckons us to hunt for conkers. If we're lucky, there are still some tomatoes ripening on the vine that will make a perfect lunch grilled with toast and a fresh poached egg. But the rewards of this season are so much more than the weather; it's harvest time and we can enjoy the fruits of our labour. The tiny seeds that we tucked up safely in the soil back in April will be proudly presenting their trophies. A rainbow of squashes, melons, peppers and beans are harvested from the raised beds, our vine is heavy with grapes, the plums are fit to burst and the apples are ready to press. But as the trees and bushes are heavy with fruit, there's no time for us to just sit back and admire them. We need to get busy, preserving the summer crops so that we can enjoy their sweet sustenance in the colder, darker months ahead.

While we are feeling thankful for our harvest, others are incorporating seasonal bounty into their own festivities. Around this time of year, Jewish families celebrate Rosh Hashanah, marking the new year in the Hebrew calendar. I was delighted to be invited to join in a children's event at a local synagogue recently, as they wanted a beekeeper to explain how honey was made. Honey is very symbolic in Jewish texts, and dipping apple slices into honey while praying for a sweet new year evokes all your senses as the custom is repeated faithfully in the traditional festive meal.

Being the daughter of two teachers, and then working in schools before having my own children, my life has always followed the rhythm of the academic year. I remember, as a child, going back to school in September after six long weeks of swimming in the sea, playing on the beach and riding my bike in the street. 'I hope you feel refreshed after your holiday and ready for a new term,' my headteacher would repeat at every year's welcome assembly, and I would think to myself, nope, I have to surrender again to the timetable, wear prescriptive uniform, be governed by lesson bells and fuelled by unappetizing school lunches. I longed for the freedom of the summer break, flying kites, picking fruit and generally engaging with nature. Little did I know at the time that the terms of the academic year revolve around harvest and that the seasons of school life very much follow traditional cycles of farming.

September is the beginning of a new school year, but it's the end of the growing year in the garden – a time to gather in, store safely and celebrate.

LEFT A barrowful back from the allotment.

CIDER, JELLY AND WINE – A CELEBRATION OF APPLES

Apples are a wonderful fruit, a true staple of both the British fruit bowl and autumn dessert. Those who grow their own, though, know that many are wasted as windfall, pecked at by birds, or riddled with the larvae of codling moth. Fortunately, cider isn't fussy. A few bruises and missing mouthfuls are not noticed as the fruit juice ferments. We do cut out the worst bits, but we're confident that the alcohol will kill any nasties that slip through the process.

As part of London's Victorian expansion, our house was built in 1883 on an old established orchard that would have previously supplied food for those living more centrally in the city. As a consequence, not only do we have some old apple trees in the middle of our garden, but there are also many more fruit trees in our area, some on public land, that yield a harvest too plentiful to consume. Cider is very forgiving and we happily welcome donations from any variety.

As apples are our biggest fruit crop, we've developed recipes that allow us to use three different preserving techniques from the bountiful harvest. Early in September, we select some of the best Bramley apples to peel, chop and freeze (see page 86), perfect for pies and crumbles throughout the winter. Fermentation (see page 106) is the preserving method in cider-making and apple wine; and making jelly from the leftover pulp results in one of our favourite preserves apple and chilli jelly (see page 158).

That first year at Hen Corner, Andy began to research how to build a cider press using a car jack and a homemade wooden frame. Two jacks later, and nearly losing the fingers of a neighbour, we decided to buy a spindle press for the following year, complete with a hand-turned crusher to prepare the apples for juicing. Now that our annual cider-making has become a regular community event, we hire in a bigger press and an electric scratter, which pulps the chopped apples ready for the press. We gather friends, some armed with extra apples, to wash, chop, crush and press the fruit; working as a team, we can extract around 85 litres/150 pints of perfumed juice to share out and take home in demijohns as it begins its journey to cider.

FROM APPLE JUICE TO CIDER – AND THAT LEFTOVER PULP

Once the apple juice has been extracted and collected into demijohns, the wild yeasts from the apples begin to ferment as they feast on the natural fruit sugars. Airlock valves are filled with water that allow carbon dioxide to escape without the cider oxidizing, and within hours the regular bubbling indicates that all is on track.

Around six weeks later, when the bubbles have all but stopped, we syphon off the cider into a clean demijohn and add a couple of tablespoons of sugar for a second ferment. The bubbles start passing through the airlock again and, after another six weeks, it's ready to bottle. We reuse bottles that we've saved throughout the year, ideally ones that are designed to take carbonated drinks as bubbles can build up pressure if you want a sparkle to your cider. We wash the bottles in hot soapy water and rinse through with sterilizing solution followed by clean hot water. For a 75cl/1¼ pint bottle, we sprinkle in a teaspoon of sugar and top up with the syphoned cider before sealing and giving a quick swirl to mix in the sugar (any size bottle will do; just adjust the sugar accordingly). It's important to syphon the cider from the demijohn with a length of tubing as there will be a layer of dead yeast in the bottom that you won't want in the final bottle. After another four weeks, the cider will be ready to drink. Time to chill it and get the glasses ready!

More recently, some of our cider-making friends have hosted a competition in winter to taste each other's brew in order to crown a champion; we've been delighted to join the party, especially as we've brought home the winner's trophy on more than one occasion.

Back to Cider Sunday . . . Once the apple juice is safely stored in fermentation containers, the press is washed ready to return and I get busy in the kitchen using the precious pulp that is the by-product of the process. I'm a firm believer in not wasting food, so if we can get two or three recipes from the same fruit it's more than worth the extra work.

I'm often asked whether we can make chutney or pie filling from the leftover apple pulp. The short answer is no, because of the chopped-up core, peel and stalks that are mixed in with it. However, any recipe that requires a liquor or stock from chopped apples can be adapted to use this valuable by-product as the very bits that we don't want to eat – the core and pips – are the source of pectin, very much needed in jelly-making. And, while the juice has been squeezed out of the fruit, there's still lots of flavour that can be extracted.

OPPOSITE Apple press ready to start.
RIGHT Our beautiful apples ready to pick.

APPLE AND CHILLI JELLY

We've been making this jelly every year since we moved to Hen Corner. It's become so popular that it's loved by locals, favoured by friends and requested by relatives. We even took it to the nation on a television show that profiled small artisan food producers. This versatile preserve, cheap and easy to make (all you need is a good supply of apples) is both a great accompaniment to food and an ingredient in its own right.

It's lovely with cheese or ham, great in marinades and glazes, perfect for super-sticky sesame sausages and even as a topping for ice cream – the list is endless.

MAKES 6 X 340G/12OZ JARS

INGREDIENTS

2kg/4lb apples (we use the leftover pulp from cider-making)

1kg/2lb granulated sugar

juice of 2 lemons

1 tbsp chilli flakes, or to taste

METHOD

1. Chop the apples into chunks about the size of a walnut. Do not peel or core.

2. Place in a large pan and cover generously with water. Bring to the boil and simmer gently for 2 hours.

3. Strain overnight through a jelly bag or suspended tea towel (tying the corners together and hanging from a cupboard door handle), carefully catching the precious apple pectin stock in a bowl below.

4. In a large preserving pan, combine 1 litre/2 pints of the apple pectin stock with the sugar, chilli flakes and lemon juice.

5. Stir over a low heat until all the sugar has dissolved, then bring to the boil and bubble rapidly at a high foaming boil for 10–15 minutes, until the setting point of 105°C/221°F is reached.

6. Turn off the heat and leave the jelly to stand in the pan for 10 minutes to cool a little and allow the chilli flakes to settle.

7. Stir and pot into hot sterilized jars (see instructions for sterilizing on page 92), then seal and label.

Adaptations

This jelly is also the perfect base for other flavours. We often substitute the chilli for preserved ginger pieces, fresh berries, rhubarb or herbs. Adding boiled sliced citrus fruit peel makes an easy marmalade.

FRUIT WINE

Most of us are familiar with wine made from grapes, but wine made from other fruit is a whole new field of home fermentation to enjoy. Recipes have been developed over the years to make use of large gluts of harvested fruit – especially popular in the days before large chest freezers. Many recipes start by soaking the fruit in a hot water solution that draws the valuable flavours and sugars from the fruit into the water, after which sugar, yeast and a few other ingredients for body and structure are added before fermentation begins and the magic starts.

If you want to try something more unusual, or use another crop you have in excess, many vegetables have also been incorporated into wine recipes. I've tried a very nice parsnip wine and I'm sure that both carrot and beetroot would work well too.

We are rather proud of our apple wine, which again uses the by-product pulp from our cider-making. It's quite a different drink to cider, more like a dessert wine, which can be enjoyed as an aperitif or as a cool spritzer when you add sparkling water and serve over ice.

ABOVE Demijohns with airlock valves and corks.

APPLE WINE

We serve this with lunch on many of our courses; it also makes a delightful gift.

MAKES A WHOLE GALLON

INGREDIENTS

1 large bucketful of chopped apples or pulp from cider pressings – approximately 5kg/10lb

1kg/2lb granulated sugar

250g/8oz chopped raisins

juice of 2 lemons

100ml/3½fl oz brewed black tea

wine yeast (for 3.8 litres/ 1 gallon of wine – follow packet instructions)

METHOD

1. Make sure your bucket has been sterilized (rinse it out with sterilizing solution). Top up the bucket of chopped apples with boiling water, cover with a lid or tea towel and leave to stand at room temperature for 7 days.

2. Remove and discard or compost the floating apple pieces.

3. Syphon the apple-infused water into a sterilized demijohn (rinse it out with sterilizing solution) and add all the other ingredients, taking care to read the manufacturer's instructions for the yeast.

4. Stir the contents of the demijohn and seal with an airlock valve (with water in the top to avoid oxygen spoiling the wine).

5. The wine will start fermenting very soon and you'll see and hear the bubbles of carbon dioxide as they rush up the airlock valve to escape. A gentle sideways turn of the demijohn each day will help keep the contents mixed, ensuring you keep all the lovely flavours mingled together.

6. After 3–4 weeks, when the bubbles have stopped, syphon the wine out of the demijohn into another sterilized demijohn, topping up to 3.8 litres/1 gallon with lukewarm water. Seal with an airlock valve.

7. After another 3–4 weeks, syphon the wine into sterilized bottles (see instructions for sterilizing bottles on page 157) and seal with corks or lids.

8. Serve chilled and enjoy.

CHRISTMAS

I always see Christmas as the mountaintop of the year, being able to look back with thankfulness at all that has gone before and, with just a few steps across to New Year, to look forward with hope towards all that is to come. If we stay with that analogy, spring can run away with us down towards the suntrap valley of summer, before we climb through the harvests of autumn up to the new mountaintop of the following year.

I love Christmas and as it's a key part of life here at Hen Corner, with special events to organize and festive treats to make in the bakery, I have most of my plans in place by mid-September.

I've always liked making food from scratch, so pickles and preserves will already be stashed away. Along with anyone who wants to join us, we make our Christmas puddings on Stir-Up Sunday (the last Sunday before Advent), when traditional church services will recite the prayer:

Stir up, O Lord, the wills of your faithful people, that they bring forth the fruit of good works, and may be richly rewarded, through Jesus Christ Our Lord. Amen

Upon hearing this, many in the congregation rush home thinking, 'Stir-up, fruit, it's Christmas pudding time!'

Christmas pudding used to be a great way to sort out the larder and mix together a variety of dried fruits – the leftovers from previous recipes – all mixed together and soaked in whatever spirit we had to spare. One year I wanted to make a figgy pudding, as in the song 'We Wish You a Merry Christmas', so figs were plentiful in that year's mix. Then there was the year that I was asked to demonstrate my Christmas pudding on stage at a large festive event. If the audience was going to taste the pudding each day, I couldn't be carefree in my quantities: I needed to have a definitive recipe and stick to it. So, dates, apricots and cranberries became the complement to the traditional vine fruits and mixed peel, and spiced rum was my tipple of choice. At the show, I worked with several other chefs, some very well known, yet there seemed to be little enthusiasm for my pudding, especially from the international chefs. 'I don't like Christmas pudding, we have tiramisu,' said the Italian chef, 'No, it's too stodgy; I make a huge croquembouche,' said the French chef. But I soon discovered that backstage my leftover pudding was being not only enjoyed but recommended and I wondered if it was nicer than they'd expected. Chatting to the hosting chef, he told me that he really liked it and that it wasn't

LEFT Our homemade wreath on the front door.

like any Christmas pudding he'd had before. So I decided to enter it into a competition and was delighted with the Great Taste Award it won.

Ideally, I'll make the Christmas cakes in October, allowing time for me to cover them with almond paste and, at a later stage, icing. Homemade Christmas crackers can be made ahead as well. I cut large squares of festive fabric with pinking shears for zigzag edges that won't fray and wrap them around cardboard tubes secured with ribbon bows. Little gifts inside could be homemade chocolates or biscuits; I've also made beeswax candles and lip balms, and Andy made little leather keyrings. I take great delight in choosing the funniest festive jokes and include folded-up paper hats. If you want your crackers to be really authentic, you can buy paper 'snaps' to thread through the middle – giving you the *bang* as you pull from both ends.

When it comes to giving gifts, I'm acutely aware that we have so much stuff. I love the advice to parents: something to wear, something to read, something they want, something they need. And unless there is something that a family member has specifically asked for, I love to give homemade gifts, hampers of goodies, consumables and experiences. Theatre tickets, a leather-working course, a chocolate-making class and other such adventures are all gifts that we have enjoyed in recent years. Let's try to go light on buying things that aren't needed and potentially generating waste.

With the lists for food and gifts in hand, as December begins, so does the countdown.

In line with midwinter traditions, we bring the outside into the warm with Christmas trees, garlands and wreaths. Made of evergreen trees and foliage, these are reminders that not all is dead in the cold dark days and that spring will surely come again.

The wreath that we hang on our front door, made with yew trimmings and decorated with holly, ivy and ribbons, is round like a wedding ring – a symbol of unending love.

OPPOSITE, LEFT
Homemade
reusable linen
crackers.
OPPOSITE, RIGHT
Natural
decorations.
RIGHT A hamper of
homemade
preserves makes a
lovely gift.
BELOW Making a
garland for the
fireplace.

The tall tree points to heaven and the slowly burning Advent candle counts down the days as we retell the story of the first Christmas.

I love Christmas carols; I love singing songs of hope and promise with neighbours in the street. I love mulled wine too, and mince pies – especially if they are homemade – and I look forward to gathering my family together with those we love.

As the big day approaches, I try to savour every moment, making time for mindfulness at the busiest time of the year. As I write Christmas cards, I pray for those I haven't seen that year; we take baked treats to neighbours and look out for those on their own.

On Christmas Eve, once the gifts have been wrapped, I walk down to the allotment to harvest veg for the feast: potatoes and parsnips, carrots and sprouts are faithfully waiting for me.

Midnight Communion ushers in the celebrations and the house is full of feasting and fun.

As we reach the end of another year, let's count our blessings – literally write them down – and look forward with ambition to learn more skills, make new friends and live more sustainably.

ABOVE Sweet parsnips ready for roasting

CHRISTMAS PUDDING

This recipe is nice and flexible, allowing you to incorporate your favourite flavours . . .

SERVES 8

INGREDIENTS

350g/12oz dried fruit (I use sultanas, dates, apricots, cranberries, mixed peel)

300ml/10fl oz spirit or liqueur (I use spiced rum)

175g/6oz fresh breadcrumbs

140g/5oz butter

40g/1½oz self-raising flour

195g/7oz dark brown sugar

1 egg, beaten

2 tsp ground mixed spice

METHOD

1. Place the fruit and alcohol in a small saucepan, warm gently for 5 minutes, then allow to steep while you prepare your other ingredients and the basin.

2. In a large bowl, cream together the butter and sugar. Add the egg, then the flour and the mixed spice.

3. Remove the saucepan from the heat and stir the breadcrumbs into the fruit mixture, then add to the mixture in the bowl and stir until everything is well combined.

4. Line a 1 litre/1¾ pint pudding basin with baking parchment and tightly fill with the pudding mix.

5. Cover the basin with a double layer of pleated parchment, then secure with string around the rim and across the top as a handle.

6. Steam for 5–6 hours, or overnight in a slow cooker, with water up to the ridge of the basin.

7. Allow to cool, then store in a cool, dry place.

8. Before serving, steam again for 1 hour to ensure it's well heated through.

CONVERSION CHARTS

VOLUME

Metric	Imperial	US cups
250ml	8fl oz	1 cup
180ml	6fl oz	¾ cup
150ml	5fl oz	⅔ cup
120ml	4fl oz	½ cup
75ml	2½fl oz	⅓ cup
60ml	2fl oz	¼ cup
30ml	1fl oz	⅛ cup
15ml	½fl oz	1 tablespoon

SPOONS

1 dessertspoon = 2 teaspoons
1 tablespoon = 3 teaspoons

1 teaspoon	5ml
2 teaspoons	10ml
1 tablespoon	15ml
2 tablespoons	30ml
3 tablespoons	45ml
4 tablespoons	60ml
5 tablespoons	75ml
6 tablespoons	90ml
7 tablespoons	105ml

Tablespoons can be easily used to convert dry (and wet) ingredients to/from US cups.

TABLESPOONS TO US CUPS

1 tablespoon	¹/₁₆ cup
2 tablespoons	⅛ cup
8 tablespoons	½ cup
10 tablespoons	⅔ cup
12 tablespoons	¾ cup
16 tablespoons	1 cup

WEIGHT

Metric	Imperial
½oz	15g
1oz	30g
2oz	60g
3oz	90g
4oz	110g
5oz	140g
6oz	170g
7oz	200g
8oz	225g
9oz	255g
10oz	280g
11oz	310g
12oz	340g
13oz	370g
14oz	400g
15oz	425g
1lb	450g

OVEN CONVERSIONS

Degrees Celsius	Degrees Fahrenheit	Gas stove mark
110°C	225°F	¼ Mark
120°C	250°F	½ Mark
135°C	275°F	1 Mark
150°C	300°F	2 Marks
165°C	325°F	3 Marks
175°C	350°F	4 Marks
190°C	375°F	5 Marks
205°C	400°F	6 Marks
220°C	425°F	7 Marks
230°C	450°F	8 Marks
245°C	475°F	9 Marks

WEIGHT OF COMMON INGREDIENTS

Keep in mind a cup of butter weighs much more than a cup of flour.
Some common ingredient conversions are as follows.

PLAIN FLOUR AND ICING SUGAR

US cups	Metric	Imperial
⅛ cup	15g	½oz
¼ cup	30g	1oz
⅓ cup	40g	1½oz
½ cup	65g	2¼oz
⅔ cup	85g	3oz
¾ cup	95g	3¼oz
1 cup	125g	4½oz

PORRIDGE OATS

US cups	Metric	Imperial
⅛ cup	10g	⅓oz
¼ cup	20g	¾oz
⅓ cup	30g	1oz
½ cup	45g	1½oz
¾ cup	60g	2¼oz
1 cup	85g	3oz

SUGAR (CASTER AND GRANULATED)

US cups	Metric	Imperial
⅛ cup	25g	1oz
¼ cup	50g	1¾oz
⅓ cup	70g	2¼oz
½ cup	100g	3½oz
⅔ cup	135g	4¾oz
¾ cup	150g	5¼oz
1 cup	200g	7oz

BUTTER AND MARGARINE

US cups	Metric	Imperial
⅛ cup	25g	1oz
¼ cup	60g	2oz
⅓ cup	75g	3oz
½ cup	110g	4oz
⅔ cup	150g	5¼oz
¾ cup	180g	6¼oz
1 cup	225g	8oz

HONEY, TREACLE AND SYRUP

US cups	Metric	Imperial
⅛ cup	45g	1½oz
¼ cup	85g	3oz
⅓ cup	110g	4oz
½ cup	170g	6oz
⅔ cup	225g	8oz
¾ cup	250g	9oz
1 cup	340g	12oz

QUICK CONVERSIONS FROM METRIC TO CUPS

Ingredient	Metric	US cups
Chocolate chips	150g	1 cup
Cocoa powder	125g	1 cup
Chopped walnuts or pecans	125g	1 cup
Walnut or pecan halves	100g	1 cup
Dessicated coconut	75g	1 cup
Baking powder	15g	1 tablespoon
Salt	18g	1 tablespoon
Grated Cheddar cheese	120g	1 cup
Grated Parmesan cheese	80g	1 cup

INFORM

Living the Good Life is an ongoing journey that's made much easier with like-minded companions. Over the years, we've been encouraged and inspired by organizations, both local and national, and numerous books.

BOOKS FOR GENERAL INSPIRATION

Hickman, L. (2005), *A Life Stripped Bare*, Eden Project Books.
Kingsolver, B. (2007), *Animal, Vegetable, Miracle*, HarperCollins.
Merrett, P. (2008), *Using the Plot: Tales of an Allotment Chef*, Collins.

MAKE

ORGANIZATIONS AND SUPPLIERS

Bread Angels is a network of people up and down the land running their own bakeries, teaching people how to make bread and how to set up their very own bakeries.
breadangels.com

Moorlands Cheesemakers Ltd Suppliers of cheesemaking equipment.
cheesemaking.co.uk

Real Bread Campaign Finding and sharing ways to make bread better for us, our communities and our planet.
sustainweb.org/realbread

World Marmalade Awards The world's premier marmalade competition!
dalemain.com/marmalade-awards

BOOKS

Bertinet, R. (2007), *Dough: Simple Contemporary Bread*, Kyle Books.
Carroll, R. (2002), *Home Cheese Making: Recipes for 75 Homemade Cheeses*. Storey Publishing.
Fearnley-Whittingstall, H. (2004), *The River Cottage Meat Book*, Hodder & Stoughton.

GROW

ORGANIZATIONS AND SUPPLIERS

Cultivate London is an independent social enterprise and charity working to improve health and well-being across London by encouraging and supporting local communities to engage with their green spaces.
cultivatelondon.org

Royal Horticultural Society is the UK's leading gardening charity, aiming to enrich everyone's life through plants and make the UK a greener and more beautiful place – committed to inspiring everyone to grow.
rhs.org.uk

BOOKS

Liebreich, K., Wagner, J. and Wenland, A. (2009), *The Family Kitchen Garden*, Frances Lincoln.
Royal Horticultural Society (2015), *Allotment Handbook & Planner: What to do to Get the Most from Your Plot*, Octopus Publishing Group.
Whittingham, J. (2007), *Vegetables in a Small Garden: Simple Steps to Success*, Dorling Kindersley.
Wright, J. (2010), *Hedgerow*, Bloomsbury Publishing.

PRESERVE

BOOKS

Pam the Jam (2019), *The Book of Preserves*, Bloomsbury Publishing.
Katz, S. (2016), *Wild Fermentation: The Flavor, Nutrition and Craft of Live-Culture Foods*, Chelsea Green Publishing.
Wong, J. (2009), *Grow Your Own Drugs: Easy Recipes for Natural Remedies and Beauty Fixes*, Collins.

KEEP

ORGANIZATIONS AND SUPPLIERS

British Beekeepers Association educates and trains beekeepers of the future and supports vital research.
bbka.org.uk

Department for Environment Food & Rural Affairs (DEFRA) is responsible for improving and protecting the environment. Aiming to grow a green economy, sustain thriving rural communities and support world-leading food, farming and fishing industries throughout the UK.
gov.uk/government/organisations/department-for-environment-food-rural-affairs

London Beekeepers Association represents the interests of beekeepers and urban beekeeping in the central London area, providing education and advice, promoting responsible beekeeping and raising awareness of the issues affecting bees.
lbka.org.uk

National Honey Show is the UK's premier honey show, with international classes, lectures, conventions and workshops.
honeyshow.co.uk

Omlet Ltd are the designers of our chicken coop Eglus and Beehaus hives. Their website also includes brilliant guides to keeping chickens and bees.
Omlet.co.uk

South Yeo Farm East produce their own succulent rare-breed meats and delicious eggs, selling the surplus to friends, neighbours and customers. They teach others how to manage their land and animals through popular smallholding courses and actively preserve the natural environment on the farm. This is where we buy eggs for hatching and meat by the whole animal to fill our freezer.
farmerdixon.co.uk

BOOKS

Campion, A. (2001), *Bees at the Bottom of the Garden*, A & C Black.

Dawson, S. (2010), *The Self-Sufficiency Bible: From Window Boxes to Smallholdings – Hundreds of Ways to Become Self-Sufficient*, Watkins Publishing.

Omlet Ltd. (2022), *What the Cluck: The Omlet Guide to Keeping Chickens*, Inkspire.

Raymond, F. (2017), *My Tiny Home Farm*, Pavilion Books.

Seymour, J. (2007), *The Concise Guide to Self-Sufficiency*, Dorling Kindersley.

Socha, P. (2016), *The Book of Bees*, Thames & Hudson. (This one is fabulous for families)

Wyndham Lewis, S. (2018), *Planting for Honeybees: The Growers Guide to Creating a Buzz*, Quadrille.

CELEBRATE

ORGANIZATIONS AND SUPPLIERS

Abundance London supports urban regeneration and gardening projects and links up forage sites/unpicked fruit trees with people who can pick.
abundancelondon.com

The English Mistletoe Shop sells mistletoe berries for propagating at home.
englishmistletoeshop.co.uk

BOOKS

Brazier, L. (2021), *Christmas at River Cottage*, Bloomsbury Publishing.

Wright, J. (2013), *Booze*, Bloomsbury Publishing.

For more information on everything in this book, why not come to Hen Corner and have a go yourself?

Be it keeping chickens, making cheese, collecting your own honey or artisan cooking skills, we are enjoying life to the full and invite you to come and try your hand at it. Most of our courses are run as small groups, for up to 6 adults, from our family home, helping you to transfer your new skills back to your own home. For those living further away, we also have a series of online/virtual courses.
Full details can be found at HenCorner.com

INDEX

AUTHOR ACKNOWLEDGEMENTS

This book was only possible because we've *been* Living the Good Life in the City and I'm grateful to all who have supported us on this journey towards self-sufficiency. I'm so thankful for our life and that we can live it to the full; journeys are so much richer, easier and rewarding when we travel with others.

First, I thank my wonderful husband Andy, my best friend on the journey, so often the voice of reason and the person who helps me to be the best I can be.

I thank our children, James for constructing chicken coops and beehives and Macy for helping with family courses and the bakery (especially during the Covid pandemic).

We are grateful to our parents, Tricia and Michael Dickinson, Pat and Brian Ward, for nurturing us as children and cheering us on as adults, and the many friends that have joined us on the journey, especially Katie Baxter, who started the bakery with me; Rachelle Attieh, who first taught me to make cheese; also Sarah and Paul New at Charlotte's Chickens, who supply us with beautiful hybrid hens.

A big shout out to Omlet for their fabulous Eglus and Beehaus and for encouraging me to run courses that teach others.

Thank you to my Beekeeping mentors, Andy Pedley, John Chapple and Thomas Bickerdike; we never stop learning do we?

Thanks also to Elsa Pawley for inspiring us with her larger livestock; don't tell Andy, but I hope to raise our own pigs soon!

Finally, thank you to the Cider Sunday gang, all the plot holders by the canal, bakery customers and course guests who contribute so much to the rich community and ethos of Hen Corner; it's great to do this together.

It's been a privilege to tell our story and bring encouragement to others through this book; thank you to all who have helped to make this happen.

Thank you to *Country Living* magazine, who commissioned me to start telling the story through a column and by speaking at their events.

Thank you to Clare Grist Taylor, who discovered Hen Corner on Tripadvisor and, after finding out more, asked if I'd ever thought of writing a book. She joined us for Cider Sunday and became my faithful agent supporting me all the way.

Thanks to Jo Christian and Gail Lynch from Pimpernel Press for believing in the story, Anna Sanderson for expertly leading me through the process, Becky Clarke for the beautiful design, Monica Hope for her copy editing and Emma O'Bryen for her support and encouragement.

Finally, a big thanks to all the friends who generously allowed us to use their photos in the book; Pam Wade, Alison Tsang, Karen Reader, Andy Ward, Richard Ward, Josiah J. Robinson and the professional photographers who kindly gave us permission to use theirs.

PICTURE CREDITS

Rupert Fowler Back cover b, 1, 4tr, 8, 48, 56, 59r, 61–63, 112, 125, 128, 149
Juliet Murphy 2, 4br, 10–11, 110, 113, 116–117, 118–119, 142
Karen Reader Back cover tl, 58–59, 127, 131, 135, 157
Josiah J. Robinson 43
Shutterstock: alexkich 19l, baibaz 26, Stepanek Photography 27, larisa Stefanjuk 28, AnaMarques 30, La corneja artesana 31, Africa Studio 32, P Kyriakos 33, DronG 38–39, Magdanatka 50, MOLPIX 52l, Wirestock Creators 52c, HandmadePictures 52r, image17 53, Anna Puzatykh 55, Vadym Zaitsev 71, Jammy Photography 73, Steve JM Hamilton 74, A_Lein 77, marcin jucha 78, kitty 79, anna.q 84, RomarioIen 87, Piotr Debowski 92b, Melica 94, Mehmet Cetin 95, David Pimborough 97, Kolpakova Svetlana 100, neil langan 103, Microgen 107, Tony Campbell 133, Tanes Ngamsom 138–139, Mycleverway 144, INTREEGUE Photography 146, Robyn Mackenzie 147, zi3000 153, pixeldreams.eu 160, Svetlana SG 166
Alison Tsang 4tl, 6, 9, 156
Andy Ward 64–65
Rich Ward 4c, 108
Sara Ward Back cover tc, 16, 21, 25, 66, 69, 105, 120, 148, 151, 154, 159, 162, 164l, 167
Pam Wade Back cover tr, 3, 4bl, 18, 19c, 19r, 20, 22, 24, 34, 36–37, 40, 45, 46–47, 76, 80–82, 88, 90–91, 92t, 93, 98, 114, 130, 136, 141, 161, 164r, 165t, 165b
Joanna Yee Front cover, 4tc, 12, 14, 122, 176

Pimpernel Press Limited
www.pimpernelpress.com

Living the Good Life in the City
© Pimpernel Press Limited 2023
Text © Sara Ward 2023
Photographs © see p174

A catalogue record for this book is available
from the British Library.

ISBN 978-1-914902-95-6

Typeset in Bodoni Egyptian Pro and Futura TT
Printed and bound in China
by C&C Offset Printing Company Limited

9 8 7 6 5 4 3 2 1